D. R. Shackleton Bailey

Homoeoteleuton in Latin dactylic verse

Beiträge zur Altertumskunde

Herausgegeben von
Ernst Heitsch, Ludwig Koenen,
Reinhold Merkelbach, Clemens Zintzen

Band 31

Springer Fachmedien Wiesbaden GmbH

Homoeoteleuton
in Latin dactylic verse

D. R. Shackleton Bailey

Springer Fachmedien Wiesbaden GmbH 1994

Die Deutsche Bibliothek – CIP-Einheitsaufnahme

Bailey, David R. Shackleton:
Homoeoteleuton in Latin dactylic verse / D. R. Shackleton
Bailey.
(Beiträge zur Altertumskunde; Bd. 31)
ISBN 978-3-663-12170-1   ISBN 978-3-663-12169-5 (eBook)
DOI 10.1007/978-3-663-12169-5

NE: GT

Satz: Passavia Druckerei Passau

PREFACE

The data for this study have been collected, classified, counted, and numerically evaluated by the author personally. Their presentation in full allows verification and opportunity for further observation and experiment. The texts for the first two sections and for the humanists Politian, Sannazarius, and Faustus Andrelinus (the 'puritans') have been read for 'homs' at least twice, the rest once. Further examination of these latter might be expected to yield some unnoticed 'homs', perhaps to the order of an added 10%, but nearly all would be in the 'omega' category, mostly 'δ^1s', merely reinforcing the evidence already assembled.

An article 'Homoeoteleuton in non-dactylic Latin verse' (*Riv. di Fil.* 120 (1992). 61–71) was submitted for publication before work began on the present study, of which it is wholly independent. Two important differences emerge. First, the incidence of 'hom' in classical dactylic verse with some exceptions (notably Horace, early and late Ovid, Germanicus, and Manilius) is far lower. A superficial observation indicates that the same holds good for Greek verse. Secondly, non-dactylic verse seems to take no special account of the 'alpha' category (noun plus attribute), so that its importance in dactylic verse came as a surprise.

Ann Arbor, 1993   D. R. Shackleton Bailey

CONTENTS

INTRODUCTORY

Lennart Håkanson's article 'Homoeoteleuton in Latin dactylic poetry' (*Harvard Stud. in Cl. Phil.* 86 [1982]. 87–115) was the first serious investigation of a curious phenomenon.[1] But basing his classification on phonetics, he missed the main point, or at any rate failed to appreciate it properly: i.e., that most classical Latin poets from Catullus on (and some post-classical ones) avoided, not so much homoeoteleuton (henceforth, apologetically, 'hom'), as the particular variety of it that occurs in juxtapositions of noun and attribute. This the data to be presented in this study will demonstrate. First, however, the term 'hom' has to be defined.

'Hom', then, in the present context is juxtaposition in the same verse of words ending in the same long vowel or long vowel plus consonant (one or two) or diphthong or nasalized vowel[2], vowels lengthened by position counting as long.

The following limitations apply:

(a) Elision of either ending nullifies the hom.

(b) When either word is followed by enclitic *que, ve,* or *ne,* there is no hom.[3]

(c) Vowels of differing quantity, one of them lengthened by position (e.g. *ignis perpetuis*), do not qualify.[4]

(d) Repeated words do not qualify.[5]

So far I follow Håkanson. Add that whereas Ennius has examples of hom in dispondaic or quadrisyllabic hexameter endings which can be admitted with the rest, Catullus in his major poems, Virgil in his *Eclogues* and *Georgics,*

---

[1] See pp. 88 f. thereof. E. Norden's Appendix IV to his commentary on *Aen.* VI is fumbling in the dark.

[2] As for metrically short vowels, ignored in this study, see Håkanson, 93–99. I doubt that further enquiry on this track would be profitable.

[3] See Håkanson, 90.

[4] Doubtful forms, i.e. *cui (cuï, quoî),* conjunctive *cum (quom),* third-declensional plurals in *-es (-is, -eis),* and the *e(n)s* ending as in *totie(n)s* seem best ignored.

[5] Including a type beloved of Ovid, as in *Met.* 5.567 *cum matre est totidem, totidem cum coniuge menses.* Håkanson adds 'cases of *-us -eus* such as *Thracius Orpheus,* which can hardly be regarded as homoeoteleuton.' Clearly they cannot: vowel and diphthong do not match.

and Ovid, all of whom shun noun-plus-attribute homs in general, do not baulk at endings like *Actaeo Aracyntho* or *purpureo narcisso;* rather they appear to affect them. Catullus and his fellow 'neoterics', Cicero's σπονδειάζοντες, notoriously favoured the dispondaic ending, and from Catullus' practice the same can be assumed for the quadrisyllabic. The 408 verses of *Peleus and Thetis* harbour thirty of one or the other, of which eight are noun-plus-attribute combinations, all of them including a Greek name or word of Greek extraction. Three are homs:

> 96   quaeque regis Galgos quaeque *Idalium frondosum*
> 141   sed conubia laeta, sed *optatos hymenaeos*
> 252   cum thiaso Satyrorum et *Nysigenis Silenis*

Since there are no other noun-plus-attribute homs in the poem, it can be hypothesized that poets who avoided them in general were not averse to this particular type, with the proviso that the combination contain a name, usually Greek, or a word of Greek origin, often (always in Ovid's case) along with hiatus.

Virgil supports. His *Eclogues* (829 lines) and *Georgics* (2186 lines), in both of which noun-plus-attribute homs are conspicuously rare, contain respectively nine[6] and fourteen dispondaic or quadrisyllabic endings. Six and eight respectively are noun-plus-attribute combinations, of which the following are homs:

> *Ecl.* 2.24   Amphion Dircaeus in *Actaeo Aracyntho*
> *Ecl.* 5.38   pro molli viola, pro *purpureo narcisso*
> *Ecl.* 7.53   stant et iuniperi et *castaneae hirsutae*
> *Ecl.* 10.12   ulla moram fecere, neque *Aonie Aganippe*
> *Georg.* 2.84   nec salici lotoque neque *Idaeis cyparissis*
> *Georg.* 4.183   et pinguem tiliam et *ferrugineos hyacinthos*[7]

The *Aeneid*, in which to be sure noun-plus-attribute homs are comparatively common, has the following:

> 3.74   Nereidum matri et *Neptuno Aegaeo*
> 3.553   Caulonisque arces et *navifragum Scylaceum*
> 4.316   per conubia nostra, per *inceptos hymenaeos*
> 7.631   Ardea Crustumerique et *turrigerae Antemnae*
> 11.31   servabat senior, qui *Parrhasio Evandro*
> 12.419   ambrosiae sucos et *odoriferam panaceam*

---

[6] Not counting the self-quotation in *Ecl.* 5.87.

[7] With these one is tempted to associate *Georg.* 1.138 *Pleiadas Hyadas claramque Lycaonis Arcton,* also Prop. 3.7.49 *sed Thyio thalamo aut Oricia terebintho.*

Propertius, apart from 3.7.49, and Tibullus have no such lines, but Ovid was rather partial to them in his pre-exilic period:

Her. 4.99	arsit et Oenides in *Maenalia Atalanta*
Her. 9.87	ut Tegeaeus aper *cupressifero Erymantho*
Her. 9.141	semivir occubuit in *lotifero Eueno*
Ars 2.185	quid fuit asperius *Nonacrina Atalanta*
Ars 3.13	si scelere Oeclides *Talaioniae Eryphylae*
Met. 2.244	et celer Ismenos cum *Phegiaco Erymantho*
Met. 3.184	nubibus esse solet aut *purpureae Aurorae*
Met. 4.535	iactari quos cernis in *Ionio immenso*
Met. 5.409	est medium Cyanes et *Pisaeae Arethusae*
Met. 8.310	cumque Pheretiade et *Hyanteo Iolao*
Met. 8.315	Penelopaeque socer cum *Parrhasio Ancaeo*
Met. 11.93	orgia tradiderat cum *Cecropio Eumolpo*
Fast. 2.43	Amphiareiades *Naupactoo Acheloo*

Germanicus ends two lines (7 and 289) with *gelidi Capricorni*, recurring in Avien. *Arat.* 56 (*gelido Capricorno*) and Auson. *Ecl.* 15.10. Poets of the silver age employ such endings very sparingly, sometimes in evident deference to Virgil or Ovid:

Manil. 5.257	pallentis violas et *purpureos hyacinthos*
Ciris 434	curalio fragili aut *electro lacrimoso*
Ciris 474	Nereidum matri et *Neptuno Aegaeo*
Colum. 10.100	nec non vel niveos vel *caeruleos hyacinthos*
Val. Fl. 7.405	abietibus tacitis aut *ignotis cyparissis*
Sil. 15.773	turba ruit, striduntque *sagittiferi goryti*
Sil. 17.79	longe coniugia ac longe *Tyrios hymenaeos*
Stat. *Theb.* 9.305	ruricolamque Gyen cum *fluctivago Ergino*
Mart. 10.12.1	Aemiliae gentes et *Apollineas Vercellas*
Juv. 6.110	sed gladiator erat; facit hoc *illos Hyacinthos*
Juv. 8.89	pone et avaritiae, miserere *inopum sociorum*
Juv. 10.229	pallida labra cibum accipiunt *digitis alienis*[8]

Occasional examples are to be found in the classicizing[9] poets of later antiquity (but not Claudian, Rutilius, or Sidonius):

Nemes. *Ecl.* 1.37	accipite hos cantus atque haec *nostro Meliboeo*
Phoenix 137	ingentes oculos credas *geminos hyacinthos*

---

[8] From Juvenal on examples sometimes lack a name or word of Greek origin.

[9] Religious poetry generally eschews such lines, though Dracontius (*Laud.* 1.121) has *lux facies rerum, dux lux* cunctis elementis and there are several with five-syllable endings in Juvencus, who is permissive of alphas in general.

Auson. *Epist.* 14.55	nodosas vestes *animantum Nerinorum*
Auson. *Ecl.* 15.10	ad tropicum pergit signum *gelidi Capricorni*
Auson. *Ecl.* 16.1	principium Iani sancit *tropicus Capricornus*
Avien. *Arat.* 56	autumni reditu cur sub *gelido Capricorno*
Avien. *Arat.* 1763	orchilos infestus si *floricomis hymenaeis*
Avien. *Descr.* 455	inter Riphaeos et *proceros Agathyrsos*
Avien. *Descr.* 777	insula Taprobane gignit *taetros elephantos*

Avienius is much addicted to dispondaic and quadrisyllabic endings[10] –
sixty in the *Aratea,* according to my count, and sixty-five in the *Descriptio.*
But only seven are noun–plus-attribute combinations, including the four
homs.[11]

I have not met with homs of this kind in medieval verse, but they recur in
neo-Latin, e.g. Politian, *Sylvae* 2.188 *lilia, sed longum stant purpurei amaranthi.*
Pontanus had a penchant for them; nine are in the 1321 lines of Grilli's
selected *Eclogues,* mostly straight Latin like 2.148 *implerunt clamore et foemineis
lamentis.*

I have not included such homs in the registers. Other types, registered
but not reckoned as homs, will be noticed presently.

The following classification will be used:

α^1 = noun[12] plus attribute (adjective or participle) in agreement.[13]
　　　Example: *Aen.* 1.225 *litoraque et* latos scopulos.
α^2 = noun with numeral adjective in agreement or with *multi, omnis, cunc-
　　　tus, totus, unus, nullus, aliquis, ille, ipse, is.*
α^3 = ablative absolute or, very rarely, gerundive construction.
　　　Example: *Aen.* 3.404 *et* positis aris *iam vota in litore solves,* Sil. 12.523
　　　tardandis Italis *corruptas igne carinas.*
α^4 = noun plus predicative attribute.
　　　Example: *Aen.* 4.321 *odere,* infensi Tyrii.
α^5 = noun plus attribute in agreement, the latter in construction with
　　　another element.
　　　Example: *Aen.* 12.629 *diva solo* fixos oculos *aversa tenebat.*

---

[10] Also with words of one or five syllables.
[11] See above.
[12] Or occasionally adjective agreeing with separated noun. This hardly ever happens
in classical verse, but Lucretius has *et* multis aliis *praebet regionibus aequor* (6.892; sim.
3.1027) and *nec iaciunt* ullum proprium *de corpore odorem* (2.846). Similarly in other cate-
gories of α, also β and γ^3.
[13] The monosyllable *di,* as in *di magni, di patrii, di superi* not included.

α^6 = noun plus passive past participle as finite verb, with or without auxiliary verb.

    Example: *Aen.* 12.269 turbati cunei *calefactaque corda tumultu.*

α^7 = noun plus epithet (hom) of like case etc. but not in agreement.

    Example: Luc. 4.244 invidia caeca *bellorum in nocte tulisset.*

β^1 = noun plus possessive adjective *meus, tuus,* or *suus* in agreement.

    Example: *Aen.* 9.543 *confixique* suis telis.

β^2 = noun plus possessive adjective *noster* or *vester* in agreement.

    Example: Hor. *A. P.* 270 *at* vestri proavi.

γ^1 = Words balancing each other antithetically or otherwise.

    Example: Ov. *Her.* 11.3 *dextra tenet* calamum, strictum *tenet altera ferrum.*

γ^2 = verbs or other parts of speech in series.

    Example: *Aen.* 9.270 *ipsum* illum (sc. *equum*), clipeum *cristasque rubentis.*

γ^3 = genitive with governing noun.

    Example: *Georg.* 1.474 *armorum sonitum.*

δ^1 = monosyllable plus word of more than one syllable.

δ^2 = *hic illic, hos illos,* et sim.

ε = two monosyllables.

ζ = two words of more than one syllable, not fitting into any of the above.

    Example: *Aen.* 5.306 *Cnosia bina* dabo levato *lucida ferro.*

The registers include all of these, but some will not count statistically as homs. γ^1 can be regarded as *exornatio* and γ^2 too seems better set aside (neither is numerically important in classical verse). Poets would naturally admit δ^2, nor do I believe that they avoided ε, the double monosyllable, for in verse of a personal nature, as Horace's *Satires* and *Epistles* and much of elegy, *me* and *te* combine freely with *de* and *ne.* Much the same applies to β^1. β^2 is not common, but an occurrence in Valerius Flaccus suggests that it was not shunned, at least not as much as α^1.[14]

In separating α^2 from α^1 I have in mind that in authors who are not particular, at least not very particular, about avoiding α^1s – Cicero, Lucretius, Horace, the *Aeneid,* Germanicus, Manilius (all of whose α^2s involve numbers), Silius, – these much outnumber α^2s. That noted, the data for Prop-

---

[14] According to Håkanson (104) 'the tendency to avoid homoeoteleuta with the preposition is notable,' but his figures (103) are too restrictive. Ovid in *Metamorphoses* has eleven examples of *a* + *-a* (eight of them in the combinations *a dextra, a laeva*) and eighteen of *pro* + *-o* (three with inversion). Comparison with prose authors may be misleading.

ertius, Tibullus, Ovid (*Metamorphoses, Ars, Tristia*) suggest that these writers were not quite so strongly averse from α^2s as from α^1s.

In view of the doubt attending the pronunciation of final nasalized vowels it has seemed advisable to separate them from the rest, calling the latter A and the former B. But so far as I can see, the data do not indicate that Bs should not be given equal consideration in this context. The relative frequency of the two types in classical verse will appear from the following table, which presents the percentage ratios of accredited B homs (i. e. α, γ^3, δ^1, and ζ) in relation to the total number of such homs (A + B), for each author, ignoring those with less than 400 verses, Ennius excepted. Total number of lines in brackets:[15]

Appendix Vergiliana		*Ex Pont.* (3194)	18
Aetna (645)	37	*Fasti* (4972)	25
Ciris (541)	25	*Heroides* 1–14 (2190)	14
Culex (414)	20	*Heroides* 16–21 (1564)	27
Calpurnius (758)	11	*Ibis* (644)	9
Catullus 62, 64–68 (700)	22	*Metamorphoses* (11995)	36
Cicero, *Aratea* (480)	32	*Tristia* (3530)	26
Columella (436)	0	Persius (650)	10
Consolatio Liviae (474)	20	Propertius	
Ennius (369)	13	I (706)	20
Germanicus (941)	20	II (1362)	12
Grattius (535)	0	III (990)	11
Horace		IV (952)	55
Satires (2103)	21	Silius (12202)	36
Epistles, A. P. (1969)	10	Statius	
Ilias Latina (1070)	36	*Achilleid* (1127)	50
Juvenal (3621)	22	*Silvae* (3313)	19
Lucan (8060)	44	*Thebaid* (9741)	30
Lucretius (7415)	44	Tibullus (1240)	35
Manilius (4258)	20	Valerius Flaccus (5591)	38
Martial (6657)	34	Virgil	
Ovid		*Aeneid* (9896)	32
Amores (2460)	28	*Eclogues* (829)	0
Ars, Med. fac., Rem. (3244)	14	*Georgics* (2188)	53

It would probably be a mistake to try to draw conclusions from these results. I find, without having any explanation to offer, that out of thirteen con-

---

[15] I have not taken account of aberrations in numbering due to interpolation or other accidents.

tinuous poems of over a thousand lines all but three (Ovid's *Ars* and *Fasti,* Manilius) have counts of from 30 to 53, whereas out of 27 shorter poems or collections of shorter poems all except Martial, Propertius IV, and Tibullus show under 31, nineteen of them under 23.[16]

Avoidance of α^{1-6}s and especially α^1s seems to have stemmed from Catullus, or perhaps the 'neoterics' generally, whose example was followed by Virgil in his *Eclogues* and *Georgics,* Tibullus, Propertius, and Ovid. Horace, whose hexameters were by his own profession[17] prose in metre, was not affected, Germanicus, Manilius, and Silius also kept more or less outside the trend. As to the *Aeneid,* see p. 100. Let the data now speak for themselves.[18] The following table of percentages of homs in ratio to total number of verses compares α^1s, α^{1-6}s, and the sum of δ^1s, ζs, α^7s, and γ^3s, which I call ω.

	α^1	α^{1-6}	ω
Appendix Vergiliana[19]			
Aetna (645)	–	0.46 (0.46 + 0)	1.07[20] (0.61 + 0.46)
Catalepton IX (64)	–	–	3.25 (3.25 + 0)
Ciris (541)	0.18 (0.18 + 0)	0.18 (0.18 + 0)	0.55 (0.37 + 0.18)
Culex (414)	0.72 (0.72 + 0)	0.97 (0.97 + 0)	1.45 (0.97 + 0.48)
Dirae (103)	0.97 (0.97 + 0)	0.97 (0.97 + 0)	2.91 (2.91 + 0)
Eleg. in Maec. (178)	–	–	1.12 (0.56 + 0.56)
Moretum (122)	–	–	3.28 (0.82 + 2.46)
Calpurnius (758)	–	–	1.45 (1.32 + 0.13)
Catullus 62, 64–68 (700)	0.14 (0.14 + 0)	0.14 (0.14 + 0)	1.00 (0.71 + 0.29)
Catullus 69–116 (314)	0.96 (0.96 + 0)	0.96 (0.96 + 0)	4.14 (2.87 + 1.27)
Cicero, *Aratea* (480)	2.08 (1.25 + 0.83)	2.70 (1.67 + 1.03)	2.64 (2.04 + 0.60)
Columella (436)	0.68 (0.68 + 0)	0.68 (0.68 + 0)	0.23 (0.23 + 0)
Consolatio Liviae (474)	–	–	2.08 (1.66 + 0.42)
Ennius, *Annals* (369)	4.61 (4.34 + 0.27)	5.96 (5.42 + 0.54)	4.33 (3.52 + 0.81)
Germanicus (941)	0.53 (0.53 + 0)	1.07 (1.07 + 0)	1.07 (0.64 + 0.43)
Grattius (535)	–	–	0.75 (0.75 + 0)

[16] Ennius and Cicero not included.

[17] In *Sat.* 1.4.39 ff.

[18] With full classified data before him the reader can form his own opinion on Håkanson's notion (106 f.) that homs were freely admitted where parts of the body, human or animal, including the words *membra* and *animus* (-*a*), were involved. In poems dealing much with physical characteristics and activities (such as fighting) such words are naturally frequent and naturally attract descriptive epithets, as a word like *populus* for instance does not.

[19] *Copa* and *Lydia* have no homs.

[20] The final digits in these totals may be out by a trifle since I have simply added the A and B percentages without recalculating.

	α^1	α^{1-6}	ω
Horace			
Satires (2103)	1.23 (0.90 + 0.33)	1.61 (1.23 + 0.38)	1.23 (1.00 + 0.23)
Epistles & A. P. (1969)	1.07 (0.97 + 0.10)	1.57 (1.37 + 0.20)	0.96 (0.86 + 0.10)
Ilias Latina (1070)	0.09 (0 + 0.09)	0.28 (0.19 + 0.09)	1.03 (0.66 + 0.37)
Juvenal (3621)	0.13 (0.08 + 0.05)	0.22 (0.17 + 0.05)	1.13 (0.88 + 0.25)
Laus Pisonis (261)	–	–	1.15 (1.15 + 0)
Lucan (8060)	0.01 (?) (0.01 (?) + 0)	0.04 (?) (0.03 (?) + 0.01)	1.25 (0.69 + 0.56)
Lucretius (7415)	0.66 (0.42 + 0.24)	1.12 (0.69 + 0.43)	1.24 (0.63 + 0.61)
Lygdamus (290)	–	–	0.37 (0.37 + 0)
Manilius (4258)	0.56 (0.44 + 0.12)	0.89 (0.73 + 0.16)	1.13 (0.68 + 0.45)
Martial (6657)	0.03 (0.03 + 0)	0.03 (0.03 + 0)	1.54 (1.00 + 0.54)
Ovid			
Amores (2460)	0.04 (0.04 + 0)	0.04 (0.04 + 0)	1.54 (1.10 + 0.44)
Ars, Rem.,			
Med. Fac. (3244)	–	0.03 (0.03 + 0)	1.15 (0.96 + 0.19)
Ex Ponto (3194)	0.03 (0.03 + 0)	0.03 (0.03 + 0)	2.04 (1.63 + 0.41)
Fasti (4972)	–	0.06 (0.04 + 0.02)	1.08 (0.82 + 0.26)
Heroides 1–14 (2190)	0.05 (0.05 + 0)	0.05 (0.05 + 0)	1.55 (1.32 + 0.23)
Heroides 15 (220)	–	–	0.09 (0.09 + 0)
Heroides 16–21 (1564)	0.13 (0.13 + 0)	0.13 (0.13 + 0)	1.80 (1.28 + 0.52)
Ibis (644)	–	–	1.86 (1.70 + 0.16)
Metamorphoses (11995)	0.07 (0.04 + 0.03)	0.13 (0.07 + 0.06)	1.15 (0.74 + 0.41)
Tristia (3530)	–	0.06 (0.06 + 0)	1.78 (1.30 + 0.48)
Halieutica (131)	–	0.75 (0.75 + 0)	–
Nux (182)	–	–	0.48 (0.48 + 0)
Panegyricus Messallae			
(211)	–	0.47 (0.47 + 0)	0.47 (0.47 + 0)
Persius (650)	0.15 (0.15 + 0)	0.30 (0.15 + 0.15)	1.23 (1.23 + 0)
Petronius,			
De bello civili (295)	0.34 (0.34 + 0)	0.34 (0.34 + 0)	1.02 (0.34 + 0.68)
Propertius			
I (706)	–	–	0.71 (0.57 + 0.14)
II (1362)	–	0.14 (0.07 + 0.07)	1.10 (0.95 + 0.15)
III (990)	0.10 (0.10 + 0)	0.30 (0.30 + 0)	0.60 (0.50 + 0.10)
IV (952)	0.11 (0.11 + 0)	0.11 (0.11 + 0)	0.84 (0.31 + 0.53)
Silius (12202)	0.17 (0.16 + 0.01)	0.35 (0.31 + 0.04)	0.89 (0.49 + 0.40)
Statius			
Achilleid (1127)	–	–	0.36 (0.18 + 0.18)
Silvae (3312)	–	–	0.74 (0.59 + 0.15)
Thebaid (9741)	0.03 (0.02 + 0.01)	0.05 (0.03 + 0.02)	0.59 (0.42 + 0.17)
Tibullus (1240)	–	0.16 (0.16 + 0)	1.21 (0.73 + 0.48)
Valerius Flaccus (5592)	–	–	0.73 (0.44 + 0.29)
Virgil			
Aeneid (9896)	0.42 (0.26 + 0.16)	0.52 (0.36 + 0.16)	0.86 (0.59 + 0.27)
Eclogues (829)	0.12 (0.12 + 0)	0.24 (0.24 + 0)	0.85 (0.85 + 0)
Georgics (2188)	0.05 (0.05 + 0)	0.14 (0.09 + 0.05)	0.54 (0.27 + 0.27)

To some extent the aversion to alpha homs must have communicated itself to omegas. The conspicuously low omega counts in Virgil's *Georgics*, Propertius I and III, Grattius, Statius, and Valerius Flaccus will be significant, as, on the other hand, will high counts in much post-classical verse, indicating that the range of 0.80% to 1.60% into which most classical poets fall will represent a degree of, probably unconscious, restraint.

For the question is inescapable, did the poets know what they were doing? External evidence that they did appears to be lacking. Hard as it may seem to accept that this feature of the 'Catullan revolution' was not present to the poet's own consciousness and that of those who followed suit, the analogy of Cicero's rhythmical practice, of which he himself, as his *Orator* discloses, was at least partly unaware, recommends the possibility. How else are we to account for the rare, occasional appearance of α^1s (like 'bad' clausulae in Cicero) in poets who normally bar them? – a phenomenon which, as will be seen, recurs in the most classically accomplished versifiers of the Renaissance.

CLASSICAL AND PRECLASSICAL

APPENDIX VERGILIANA

Aetna (645 lines)

	A		B	
α^2	3	0.46	–	
δ^1	3	0.46	2	0.31
δ^2	1	0.15	–	
ζ	1	0.15	1	0.15

Catalepton IX (64 lines)

	A		B	
δ^1	2	3.25	–	
ε	1	1.59	–	

Ciris (541 lines)

	A		B	
α^1	1	0.18	–	
γ^1	1	0.18	–	
δ^1	1	0.18	1	0.18
ε	1	0.18	–	
ζ	1	0.18	–	

Culex (414 lines)

	A		B	
α^1	3	0.72	–	
α^2	1	0.24	–	
α^7	1	0.24	–	
δ^1	1	0.24	1	0.24
ζ	2	0.48	1	0.24

Dirae (103 lines)

	A		B	
α^1	1	0.97	–	
β^1	1	0.97	–	
δ^1	2	1.94	–	
ζ	1	0.97	–	

Elegiae in Maecenatem (178 lines)

	A		B	
γ^1	1	0.56	–	
δ^1	–		1	0.56
ζ	1	0.56	–	

Moretum (122 lines)[1]

	A		B	
δ^1	–		3	2.46
ζ	1	0.82	–	

[1] *Copa* (58 lines) and *Lydia* (80 lines) have no homs.

Aetna

α^2 A

220   unde *ipsi venti,* quae res incendia pascit
307   *principiis aliis* credas consurgere ventos
630   illis *divitiae solae* materque paterque

δ^1 A

36   discrepat *a prima* facies haec altera vatum
132   quod *si praecipiti* conduntur flumina terra
536   quod si *quis lapidis* miratur fusile robur

δ^1 B

259   *tum demum* vilesque iacent inopesque relictae
464   *tum pavidum* fugere et sacris concedere rebus

δ^2 A

447   *huc illuc* ageret ventos et pasceret ignes

ζ A

481   *exustus penitus* venis subit altius umor

ζ B

224   non oculis *solum pecudum* miranda tueri

Catalepton IX

δ^1 A

32   saepe *rubro pro* qua sanguine fluxit humus
37   *illo quo* primum dominatus Roma superbos

ε A

10   quid *de te* possim scribere, quidve tibi

Ciris

α^1 A

238   ille *Arabae Myrrhae* quondam qui cepit ocellos

γ^1 A

352   Hesperium *vitant, optant* ardescere Eoum

$\delta^1 A$

52   hanc pro *purpureo* poenam scelerata capillo

$\delta^1 B$

248   milia *visuram, quam* te tam tristibus istis

εA

408   vos † o numantana † *si qui* de gente venitis

ζA

373   despue ter, *virgo: numero* deus impare gaudet

Culex

$\alpha^1 A$

328   et iam *Strymonii Rhesi* victorque Dolonis
340   ne quisquam *propriae Fortunae* munere dives
398   conserit, *assiduae curae* memor, hic et acanthos

$\alpha^2 A$

159   anxius *insidiis nullis,* sed lentus in herbis

α^7

165   mersus ut in *limo magno* subsideret aestu

$\delta^1 A$

110   Delia diva, *tuo, quo* quondam victa furore

$\delta^1 B$

376   ergo *iam causam* mortis, iam dicere vitae

ζA

1   lusimus, *Octavi, gracili* modulante Thalia
74   pallentemque *sibi viridi* cum gramine lucens

ζB

95   fontis *Hamadryadum, quarum* non divite cultu

Dirae

α¹ A

74   † coculet † *arguti grylli* cava garrula rana

β¹ A

79   cum delapsa *meos agros* pervenerit unda

δ¹ A

48   *undae, quae* vestris pulsatis litora lymphis
81   advena, *civili qui* semper crimine crevit

ζ A

54   tristius hoc, memini, revocasti, Battare, carmen[2]

Elegiae in Maecenatem

γ¹ A

169   arbiter ipse *fui, volui* quod contigit esse

δ¹ B

67   mollius es solito *mecum tum* multa locutus

ζ A

125   illius aptus *eras roseas* adiungere bigas

Moretum

δ¹ B

20   *quam fixam* paries illos servabat in usus
92   singula *tum capitum* nodoso corpore nudat
114   *tum demum* digitis mortaria tota duobus

ζ A

52   Simulus *interea vacua* non cessat in hora

[2] Sim. 71.

CALPURNIUS (758 lines)

	A		B	
γ^1	1	0.13	–	–
δ^1	8	1.06	1	0.13
ϵ	2	0.26	–	
ζ	2	0.26	–	

γ^1

4.62   Tityrus hanc *habuit, cecinit* qui primus in istis

δ^1 A

3.11   quae sibi, nam *memini, si* quando solus abesses
3.97   nam bonus *a dextra* fecit mihi Tityrus omen
4.28   praeter *ab his scopulis* ventosa remurmurat echo
4.76   hos potius, magis *hos calamos*[3] sectare: canales
6.16   qui posset *dici, si* non cantaret, Apollo
7.4   o piger, *o duro* non mollior axe, Lycota
7.39   cum mihi iam senior, *lateri qui* forte sinistro
7.51   et coit in rotulum, *tereti qui* lubricus axe

δ^1 B

6.49   aspice *quam timeam:* genus est, ut scitis, equarum

ϵ A

2.80   et *nos, quos* etiam praetorrida munerat aestas
6.88   efficerem *ne te* quisquam tibi turpior esset

ζ A

2.74   semper *olus metimus,* nec bruma nec impedit aestas
4.160   tum mihi talis *eris, qualis* qui dulce sonantem

---

[3] V.1. *calamos, magis hos.*

CATULLUS

	A		B	
62, 64–68 (700 lines)				
α^1	1	0.14	–	
β^1	1	0.14	–	
δ^1	5	0.71	1	0.14
δ^2	1	0.14	–	
ϵ	1	0.14	–	
ζ	–	–	1	0.14
69–116 (314 lines)				
α^1	3	0.96	–	
γ^1	2	0.63	1	0.31
δ^1	7	2.23	2	0.63
ϵ	1	0.31	–	
ζ	2	0.63	2	0.63

To him, or to him and his associates, is probably to be ascribed the strong dislike of alpha homs, accompanied at least in some cases by some degree of aversion to omegas, which became a feature of most classical and some postclassical dactylic Latin verse. The difference between his longer, 'serious' poems and the shorter dactylic pieces is unlikely to be due to accident and paucity of material.

In 64.122 the standard reading ⟨*venerit,*⟩ *aut ut eam devinctam lumina somno,* with Lachmann's supplement, can now be ruled out. Laetus' *aut ut eam* ⟨*placido*⟩ *devinctam lumina somno,* quoted in Mynors' apparatus, is unobjectionable.

62, 64–68

α^1 A

66.15   estne *novis nuptis* odio Venus? anne parentum

β^1 A

67.30   qui ipse *sui gnati* minxerit in gremium

δ^1 A

64.228   quod tibi *si sancti* concesserit incola Itoni
64.286   Tempe, *quae silvae* cingunt super impendentes

66.1	omnia *qui magni dispexit lumina mundi*
66.26	cognoram *a parva* virgine magnanimam
68.160	lux mea, *qua viva* vivere dulce mihi est

δ^1 B

| 64.296 | *quam quondam* silici restrictus membra catena |

δ^2 A

| 68.133 | quam circumcursans *hinc illinc* saepe Cupido |

ε A

| 64.152 | *pro quo* dilaceranda feris dabor alitibusque |

ζ B

| 66.67 | vertor in *occasum, tardum* dux ante Booten |

69–116

α^1 A

89.3	tamque *bonus patruus* tamque omnia plena puellis
106.1	cum *puero bello* praeconem qui videt esse
108.5	*effossos oculos* voret atro gutture corvus

γ^1 A

| 74.5 | quod *voluit fecit:* nam quamvis irrumet ipsum |
| 111.3 | sed *cuivis quamvis* potius succumbere par est |

γ^1 B

| 73.1 | desine de *quoquam quicquam* bene velle mereri |

δ^1 A

76.12	et *dis invitis* desinis esse miser
82.1	*Quinti, si* tibi vis oculos debere Catullum
88.1	quid facit is, *Gelli, qui* cum matre atque sorore
92.3	*quo signo?* quia sunt totidem mea: deprecor illam
96.3	*quo desiderio* veteres renovamus amores
101.2	advenio *has miseras,* frater, ad inferias
109.3	*di magni,* facite ut vere promittere possit

δ^1 B

| 99.15 | *quam quoniam poenam* misero proponis amori |
| 99.16 | *numquam iam* posthac basia surripiam |

ε A

92.2   *de me:* Lesbia me dispeream nisi amat

ζ A

77.2   *frustra? immo magno* cum pretio atque malo
78.2   *alterius, lepidus* filius alterius

ζ B

74.6   nunc *patruum, verbum* non faciet patruus
99.15   *quam quoniam poenam* misero proponis amori

CICERO

	A		B	
Aratea (480 lines)				
α^1	6	1.25	4	0.83
α^2	1	0.20	1	0.20
α^4	1	0.20	–	
α^7	2	0.40	–	
γ^1	1	0.20	–	
δ^1	3	0.60	2	0.40
ζ	5	1.00	1	0.20

The figures reflect intermediate status between Ennius and Lucretius. The seventy-eight surviving lines from *De consulatu suo* (fr. 11 Morel) contain two α^1As (31 *perpetuis signis* and 77 *anxiferas curas*), one α^1B (50 *ingentem cladem*), and no other homs of any kind. Nor are there any in the thirteen lines of the *Marius* (fr. 7) and the fifty-three from Homer (fr. 22–30), except for *his dextris fulgoribus* (fr. 25).

α^1 A

43   *Mercurius parvus manibus* quam dicitur olim
202   coeperit et *subitis auris* diduxerit Ara
256   hunc *sura laeva* Perseus umeroque sinistro
280   hunc, a *clarisonis auris* Aquilonis ad Austrum
296   hosce *aequo spatio* devinctos sustinet axis
313   et *quantos radios* iacimus de lumine nostro

$\alpha^1\,\mathrm{B}$

63   nam non *longinquum spatium* labere diurnum
187   Arcturo, *magnum spatium* supero dedit orbe
274   atque pedes *gelidum rivum* fundentis Aquari
383   atque Avis ad *summam caudam* primasque recedit

$\alpha^2\,\mathrm{A}$

153   hac *una stella* nectuntur, quam iacit ex se

$\alpha^2\,\mathrm{B}$

262   at vero *totum spatium* convestiet orbis

$\alpha^4\,\mathrm{A}$

115   *suspensos animos* arbusta ornata tenere

$\alpha^7\,\mathrm{A}$

59   corpore *semifero magno* Capricornus in orbe
305   quam sunt in *caelo divino* numine flexi

$\gamma^1\,\mathrm{A}$

306   terram *cingentes, ornantes* lumine mundum

$\delta^1\,\mathrm{A}$

96   *illae quae* fulgent luces ex ore corusco
194   *a summa* parte obscura caligine tectam
370   sed cum de *terris vis* est patefacta Leonis

$\delta'\,\mathrm{B}$

184   *Aram, quam* flatu permulcet spiritus Austri
398   nondum tota patet, *nam caudam* contegit umbra

$\zeta\,\mathrm{A}$

6   quod *soliti, simili* quia forma littera claret
43   *Mercurius parvus manibus* quam dicitur olim
110   haud *vero toto* spirans de corpore flammam
212   quam *nemo certo* donavit nomine Graium
266   hic *totus medius* circlo distinguitur; iste

$\zeta\,\mathrm{B}$

406   nam *secum medium* pandet Nepa, tollere vero

COLUMELLA

	A		B	

De cultu hortorum (436 lines)

	A		B	
α^1	3	0.68	–	
γ^2	–	1	0.23	
ζ	1	0.23	–	

A freak, so far as it goes.

α^1 A

112	alliaque *infractis spicis* et olentia late
121	rutaque *Palladiae baccae* iutura saporem
427	inter *lascivos Satyros* Panasque biformes

γ^2 B

| 210 | ver agit: hinc *hominum pecudum* volucrumque cupido |

ζ A

| 178 | nunc veniat *quamvis oculis* inimica coramble |

CONSOLATIO LIVIAE (474 lines)

	A		B	
γ^3	–		1	0.21
δ^1	6	1.64	–	
δ^2	1	0.21	–	
ε	1	0.21	–	
ζ	2	0.42	1	0.21

γ^3 B

| 339 | vix credent *tantum rerum* cepisse tot annos |

δ^1 A

32	iam mihi *pro Druso* dona ferenda meo
129	Caesaris *uxori si* talia dicere fas est
134	et vorat *hos ipsos* flamma rogusque sinus
195	Livia, non illos *pro Druso* Livia movit

463   et *quo me officio* portaverit illa iuventus
468   hoc ego *qui flendi* sum tibi causa rogo

$$\delta^2 A$$

184   *hic illic* pavidi clamque palamque dolent

$$\varepsilon A$$

299   quid referam *de te,* dignissima coniuge Druso

$$\zeta A$$

148   iamne *fui Drusi* mater, et ipse fuit
419   vix etiam *fueras paucas* vitalis in horas

$$\zeta B$$

301   par bene *compositum: iuvenum* fortissimus alter

ENNIUS

Annals (369 lines)

	A		B	
α^1	16	4.34	1	0.27
α^2	2	0.54	1	0.27
α^3	1	0.27	–	
α^6	1	0.27	–	
α^7	1	0.27	–	
β^1	1	0.27	–	
β^2	1	0.27	–	
γ^1	1	0.27	–	
γ^2	1	0.27	1	0.27
δ^1	7	1.89	1	0.27
ε	3	0.81	1	0.27
ζ^5	5	1.36	2	0.54

As in Plautus' iambics, homs run riot. Both alpha and omega counts are enormous by classical standards, especially the former.

I use the numbering in Warmington's *Remains of old Latin I* (Loeb).

$\alpha^1\,\mathrm{A}$

25	quam *prisci casci populi*[4] tenuere Latini
31	olli respondit rex *Albai Longai*[5]
161	*hastis ansatis,* concurrunt undique telis
166	⟨quam⟩ *nonis Iunis* soli Luna obstitit et nox
185	arbustum fremitu *silvai frondosai*
201	ut *pro Romano populo* prognariter armis
240	stant *sectis foliis* et amaro corpore buxum
250	deducunt habiles gladios *filo gracilento*
265	haud *doctis dictis* certantes sed maledictis
318	*rastros dentiferos* carpsit causa poliendi
349	quae neque *Dardaniis campis* potuere perire
400	nox quando *mediis signis* praecincta volabit
489	optima cum *pulchris animis* Romana iuventus
518	vincla *suis magnis animis* abrupit et inde
524	deque totondit *agros laetos* atque oppida cepit

$\alpha^1\,\mathrm{B}$

538	dum *clavum rectum* teneam navemque gubernem

$\alpha^2\,\mathrm{A}$

285	et rursus *multae fortunae* forte recumbunt
549	*omnes mortales* sese laudarier optant

$\mathrm{A}^2\,\mathrm{B}$

225	*multorum veterum,* leges divumque hominumque

$\alpha^3\,\mathrm{A}$

206	ut primum *tenebris abiectis* indalbabat

$\alpha^6\,\mathrm{A}$

109	nam vi depugnare sues *stolidi soliti* sunt

$\alpha^7\,\mathrm{A}$

214	consilio indu *foro lato*[6] sanctoque senatu

$\beta^1\,\mathrm{A}$

518	vincla *suis magnis animis* abrupit et inde

[4] Counted twice.
[5] See p. 1.
[6] Participle, not adjective.

β² A

228   ⟨heu⟩ quianam *dictis nostris* sententia flexa est

γ¹ A

384   malos *diffindunt, fiunt* tabulata falaeque

γ² A

298   *parerent observarent,* portisculu' signum

γ² B

127   *Volturnalem Palatualem Furinalem*

δ¹ A

42   *his verbis:* 'o gnata, tibi sunt ante ferendae
52   Ilia, dia nepos, *quas aerumnas* tetulisti
134   ferro se caedi quam *dictis* his toleraret
154   postquam lumina *sis oculis* bonus Ancu' reliquit
193   dono, ducite, doque volentibu' cum *magnis dis*
201   ut *pro Romano populo* prognariter armis
399   *si luci si* nox si mox si iam data sit frux

δ¹ B

24   est locus *Hesperiam quam* mortales perhibebant

ε A

50   ut *me de* caelo visas cognata parumper
217   evomeret *si qui* vellet tutoque locaret
441   hortatore bono priu' *quam iam* finibu' termo

ε B

230   in somnis vidit priu' *quam sam* discere coepit

ζ A

92   rebus *utri magni* victoria sit data regni
209   quem *nemo ferro* potuit superare nec auro
275   certare *abnueo; metuo* legionibu' labem
347   contendunt *Graecos, Graios* memorare solent sos
430   concurrunt *veluti venti* quom spiritus Austri

ζ B

229   nec *quisquam sophiam* sapientia quae perhibetur
286   *haudquaquam quemquam* semper Fortuna secuta est

GERMANICUS (941 lines)

	A		B	
α^1	5	0.53	–	
α^2	4	0.43	–	
α^4	1	0.11	–	
α^7	1	0.11	–	
β^1	1	0.11	–	
δ^1	2	0.22	2	0.22
ε	1	0.11	–	
ζ	3	0.33	2	0.22

An alpha count as high as the omega makes him one of the exceptions to
the classical norm.

In 238 the manuscript reading *unius brevior* has been defended as avoiding
the hom *unius brevius,* but cf. 536 *taurus, cuius.*

α^1 A

Aratea 58	*tempus dexterius* quae signat stella draconis
Aratea 76	et *vastos umeros,* tum cetera membra secuntur
Aratea 203	sic *magnis umeris* candet nitor. hanc media ambit
Aratea 615	et *niveus cycnus* properarit tangere fluctus
Aratea 697	non *vastos umeros,* non pectora tristia saetis

α^2 A

Aratea 383	sunt *aliae stellae;* qua caudam belua flectit
Aratea 437	quinque *aliae stellae* diversa lege feruntur
Aratea 592	mergitur in *totos umeros* Ophiuchus et anguis
Fr. IV.117	quin *alias pluvias* alia in regione notabis

α^4 A

Aratea 367	*Eridanus medius* liquidis interiacet astris

α^7 A

Fr. IV.86	et modo de *vento, gelido* modo protinus imbre

β^1 A

Aratea 635	arcus sive *suo caelo* referetur imago

δ¹ A

Aratea 506	ultima, deficiunt *nigra qua* sidera cauda
Fr. II.1	una via est *solis bis* senis lucida signis

δ¹ B

Aratea 184	Iasides etiam *caelum cum* coniuge Cepheus
Aratea 201	nec procul Andromede, *totam quam* cernere posse

ε A

Aratea 454	dividitur, binos ut *si qui* desecet arcus

ζ A

Aratea 83	impar est *manibus pondus,* nam dextera parvam
Aratea 238	sunt spatia; *unius brevius*[7] sed clarior ignis
Aratea 536	corniger hic *taurus, cuius* decepta figura

ζ B

Aratea 569	Oceani, *tantum liquidum* super aera lucet
Aratea 589	cum *primum Cancrum* Tethys emittit in auras

GRATTIUS (535 lines)

	A		B
δ¹	4 (?)	0.75	–
ε	1	0.19	–

Those who read *Naides, et Latii ⟨satyri⟩ Faunusque subibant* in 18 unwittingly present the author with a unique α¹, for which he might not thank them. *Panes* would do as well.

δ¹ A

373	quod sive *a Stygia* letum Proserpina nocte
408	at *si deformi*[8] lacerum dulcedine corpus
428	materiem largita *boni, si* vincere curent
439	auxilia et, *meriti si* nulla est noxia tanti[9]

ε A

419	laxatusque rigor. quae *te ne* cura timentem

---

[7] See above.

[8] *deformis* Ios. Wassius.

[9] *meritis (A) ... tanta* Baehrens.

HORACE

	A		B	
Satires (2103 lines)				
α^1	19	0.90	7	0.33
α^2	2	0.10	1	0.05
α^3	5	0.24	–	
α^7	1	0.05	–	–
β^1	1	0.05	–	
γ^1	3	0.14	1	0.05
δ^1	16	0.76	4	0.19
ε	11	0.52	–	
ζ	4	0.19	1	0.05
Epistles and *A. P.* (1969 lines)				
α^1	19	0.97	2	0.10
α^2	1	0.05	2	0.10
α^3	6	0.30	–	
α^5	1	0.05	–	
β^1	1	0.05	–	
β^2	1	0.05	3	0.15
γ^1	6	0.30	1	0.05
γ^2	1	0.05	–	
δ^1	16	0.81	–	
ε	10	0.51	1	0.05
ζ	1	0.05	1	0.05

Uniquely among classical poets who have left more than a thousand dactylic lines, the alpha count in Horace's hexameters exceeds the omega, and that by a wide margin. No Catullian flummery for him. His α^{1-6} + ω counts of 2.84 and 2.53 invite comparison with Lucretius' 2.34, Manilius' 2.02, Germanicus' 2.04, and the *Culex'* 2.42; far below those of Ennius (10.29), Cicero (5.34), and Catullus' *minora* (5.10).[10]

[10] These figures, taken by adding the totals for α^{1-6} and ω in the table on p. 7 f., are not absolute (see p. 7 n. 20), but they are close enough.

Satires

α¹ A

1.2.25	Maltinus *tunicis demissis* ambulat; est qui
1.2.132	*discincta tunica* fugiendum sit, pede nudo
1.3.134	*lascivi pueri;* quos tu nisi fuste coerces
1.4.128	sic *teneros animos* aliena opprobria saepe
1.5.2	*hospitio modico;* rhetor comes Heliodorus
1.5.34	Fundos *Aufidio Lusco* praetore libenter[11]
1.6.94	a *certis annis* aevum remeare peractum
1.7.17	cum *Lycio Glauco,* discedat pigrior ultro
1.8.20	*humanos animos.* has nullo perdere possum
1.9.1	ibam forte *via Sacra,* sicut meus est mos
1.9.70	*curtis Iudaeis* oppedere?' 'nulla mihi' inquam
2.2.78	*hesternis vitiis* animum quoque praegravat una
2.3.258	porrigis *irato puero* cum poma, recusat
2.4.2	ponere signa *novis praeceptis,* qualia vincent
2.5.86[12]	unctum *oleo largo nudis umeris* tulit heres
2.5.99	cum te *servitio longo* curaque levarit
2.7.82	duceris ut *nervis alienis* mobile lignum
2.8.18	*divitias miseras!* sed quis cenantibus una

α¹ B

1.4.110	Baius inops? *magnum documentum* ne patriam rem
1.4.133	*consilium proprium.* neque enim, cum lectulus aut me
1.5.36	praetextam et *latum clavum* prunaeque vatillum
1.6.25	sumere *depositum clavum* fierique tribuno
1.10.67	quamque *poetarum seniorum* turba; sed ille
2.2.97	*iratum patruum,* vicinos, te tibi iniquum
2.3.63	*errorem similem* cunctum insanire docebo

α² A

1.1.81	aut *alius casus* lecto te affixit, habes qui
1.6.103	exirem, *plures calones* atque caballi

α² B

1.3.80	si quis *eum servum* patinam qui tollere iussus

---

[11] Properly belongs in a special category, but there seems to be no other example.
[12] Counted twice.

α^3 A

1.4.22	*delatis capsis* et imagine, cum mea nemo
1.10.77	*contemptis aliis* explosa Arbuscula dixit
2.1.68	famosisve *Lupo cooperto* versibus? atqui
2.3.68	*reiecta praeda* quam praesens Mercurius fert
2.5.95	*oppositis umeris;* aurem substringe loquaci

α^7

1.10.87	compluris *alios, doctos* ego quos et amicos

β^1 A

1.2.76	immiscere. *tuo vitio* rerumne labores

γ^1 A

1.5.11	tum pueri *nautis, pueris* convicia nautae
2.3.104	si quis emat *citharas, emptas* comportet in unum
2.7.71	cum semel *effugit, reddit* se prava catenis

γ^1 B

2.3.325	mille *puellarum, puerorum* mille furores

δ^1 A

1.2.77	nil referre putas? *quare, ne* paeniteat te
1.2.90	hoc illi *recte. ne* corporis optima Lyncei
1.4.91	infesto nigris; ego *si risi* quod ineptus
1.4.99	sed tamen admiror *quo pacto*[13] iudicium illud
1.6.30	ut si qui aegrotet *quo morbo* Barrus haberi
1.6.63	quod placui tibi, *qui turpi* secernis honestum
1.9.40	et *propero quo* scis.' 'dubius sum quid faciam' inquit
1.9.55	expugnabis; et est *qui vinci* possit, eoque
2.2.35	ducit te species, *video; quo* pertinet ergo
2.3.129	incipias servosque *tuos, quos* aere pararis
2.3.248	ludere *par impar,* equitare in harundine longa
2.5.31	sperne, *domi si* gratus erit fecundave coniunx
2.5.87	scilicet *elabi si* posset mortua; credo
2.6.31	ad Maecenatem *memori si* mente recurras
2.7.110	furtiva mutat *strigili? qui* praedia vendit
2.8.21	*si memini,* Varius, cum Servilio Balatrone

---

[13] Also in 1.7.2, 2.4.8, 2.7.22, *Epist.* 1.8.13, 2.1.171. In this case, and in this case only (Lucretius excepted), I have left such repetitions out of reckoning.

δ¹ B

1.1.16	*iam faciam* quod vultis, eris tu, qui modo miles
1.4.20	usque laborantis, *dum ferrum* molliat ignis
1.10.57	quaerere num illius, *num rerum* dura negarit
2.3.317	quantane, *num tantum* (sufflans se) magna fuisset
2.3.318	'maior dimidio.' '*num tantum*?'[14] cum magis atque

ε A

1.1.14	delassare valent Fabium. *ne te* morer, audi
1.1.69	flumina-quid rides? mutato nomine *de te*
1.1.78	*ne te* compilent fugientes, hoc iuvat? horum
1.1.98	supremum tempus *ne se* penuria victus
1.1.120	iam satis est. *ne me* Crispini scrinia lippi
1.4.41	dixeris esse satis; neque *si qui* scribat, uti nos
1.4.102	atque animo, prius ut, si quid promittere *de me*
1.10.55	cum *de se* loquitur non ut maiore reprensis
2.3.31	'dum ne quid simile huic, esto ut libet.' 'o bone, *ne te*
2.6.36	*de re* communi scribae magna atque nova te
2.8.25	Nomentanus ad hoc, *qui, si* quid forte lateret

ζ A

1.1.48	forte vehas *umero, nihilo* plus accipias quam
1.1.66	sic *solitus:* '*populus* me sibilat, at mihi plaudo
1.6.45	nunc ad me *redeo libertino* patre natum
1.10.30	verba *foris malis* Canusini more bilinguis

ζ B

1.1.24	*percurram: quamquam* ridentem dicere verum

Epistles and *Ars Poetica*

α¹ A

Epist. 1.1.70	quod si me *populus Romanus* forte roget cur
Epist. 1.2.1	*Troiani belli* scriptorem, Maxime Lolli
Epist. 1.2.52	ut lippum *pictae tabulae,* fulmenta podagrum
Epist. 1.4.11	et *mundus victus* non deficiente crumina
Epist. 1.5.18	*sollicitis animis* onus eximit, addocet artis

---

[14] Reckoned as one with the preceding. The readings are in doubt.

Epist. 1.10.4	*fraternis animis,* quidquid negat alter, et alter
Epist. 1.16.38	mordear *opprobriis falsis* mutemque colores
Epist. 1.16.51	*suspectos laqueos* et opertum miluus hamum
Epist. 1.17.40	ut *parvis animis* et parvo corpore maius
Epist. 1.18.33	cum *pulchris tunicis* sumet nova consilia et spes
Epist. 1.18.35	officium, *nummos alienos* pascet, ad imum
Epist. 2.1.188	omnis ad *incertos oculos* et gaudia vana
Epist. 2.2.7	*litterulis Graecis* imbutus, idoneus arti
Epist. 2.2.201	non agimur *tumidis velis* Aquilone secundo
Ars 37	spectandum *nigris oculis* nigroque capillo
Ars 77	quis tamen *exiguos elegos* emiserit auctor
Ars 161	*imberbis*[15] *iuvenis,* tandem custode remoto
Art 325	*Romani pueri* longis rationibus assem
Ars 340	neu *pransae Lamiae* vivum puerum extrahat alvo

$$\alpha^1 \, B$$

Epist. 1.1.43	esse mala, *exiguum censum* turpemque repulsam
Ars 147	nec gemino *bellum Troianum* orditur ab ovo

$$\alpha^2 \, A$$

Epist. 1.19.41	hinc *illae lacrimae.* ʻspissis indigna theatris

$$\alpha^2 \, B$$

Epist. 1.15.14	maior *utrum populum* frumenti copia pascat
Ars 142	qui mores *hominum multorum* vidit et urbis

$$\alpha^3 \, A$$

Epist. 1.6.14	*defixis oculis* animoque et corpore torpet
Epist. 2.2.104	idem, *finitis studiis* et mente recepta
Epist. 2.2.134	et *signo laeso* non insanire lagonae
Ars 166	*conversis studiis* aetas animusque virilis
Ars 447	*traverso calamo* signum, ambitiosa recidet
Ars 3	undique *collatis membris,* ut turpiter atrum

$$\alpha^5 \, A$$

Epist. 2.2.178	saltibus *adiecti Lucani, si* metit Orcus

$$\beta^1 \, A$$

Epist. 1.18.65	consentire *suis studiis* qui crediderit te

---

[15] V.l. *imberbus.*

$$\beta^2\,A$$

Ars 270   at *vestri proavi* Plautinos et numeros et

$$\beta^2\,B$$

Epist. 1.3.36   pascitur in *vestrum reditum* votiva iuvenca
Epist. 1.4.1   Albi, *nostrorum sermonum* candide iudex
Epist. 1.14.31   nunc age, quid *nostrum concentum* dividat audi

$$\gamma^1\,A$$

Epist. 1.1.98   quod petiit *spernit, repetit* quod nuper omisit
Epist. 1.1.100   diruit *aedificat, mutat* quadrata rotundis
Epist. 1.6.12   gaudeat an *doleat, cupiat* metuatne, quid ad rem
Epist. 1.7.86   verum ubi oves *furto, morbo* periere capellae
Epist. 1.16.65   non video. nam qui *cupiet, metuet* quoque; porro
Epist. 1.17.12   te tractare *voles, accedes* siccus ad unctum

$$\gamma^1\,B$$

Epist. 2.1.257   si quantum *cuperem possem* quoque. sed neque parvum

$$\gamma^2\,A$$

Ars 430   ex oculis rorem, *saliet, tundet* pede terram

$$\delta^1\,A$$

Epist. 1.2.40   dimidium *facti, qui* coepit, habet: sapere aude
Epist. 1.6.8   quo spectanda *modo, quo* sensu credis et ore
Epist. 1.6.32   lucum ligna: *cave ne* portus occupet alter
Epist. 1.6.68   candidus *imperti; si* nil, his utere mecum
Epist. 1.12.12   miramur *si Democriti* pecus edit agellus
Epist. 1.16.15   *hae latebrae* dulces, etiam, si credis, amoenae
Epist. 1.16.26   dicat et *his verbis* vacuas permulceat auris
Epist. 1.16.46   'nec furtum feci nec *fugi' si* mihi dicit
Epist. 1.19.43   *si dixi,* 'rides' ait 'et Iovis auribus ista
Epist. 2.1.138   carmine *di superi* placantur, carmine Manes
Epist. 2.2.40   ibit *eo quo* vis qui zonam perdidit, inquit
Epist. 2.2.178   saltibus *adiecti Lucani, si* metit Orcus
Ars 327   filius *Albani: si* de quincunce remota est
Ars 431   ut *qui conducti* plorant in funere dicunt
Ars 451   offendam in nugis?' *hae nugae* seria ducent
Ars 458   *si veluti* merulis intentus decidit auceps

ε A

Epist. 1.1.105	*de te pendentis, te respicientis amici*
Epist. 1.6.24	ostendat tibi *te, ne* fidos inter amicos
Epist. 1.7.82	mercatur, *ne te* longis ambagibus ultra
Epist. 1.15.40	aiebat, '*si qui* comedunt bona, cum sit obeso
Epist. 1.16.19	sed vereor ne cui *de te* plus quam tibi credas
Epist. 1.18.58	ac *ne te* retrahas et inexcusabilis absis
Epist. 1.18.98	*ne te* semper inops agitet vexetque cupido
Epist. 1.19.26	ac *ne me* foliis ideo brevioribus ornes
Epist. 2.1.245	at neque dedecorant tua *de se* iudicia atque
Epist. 2.2.191	tollam, nec metuam quid *de me* iudicet heres

ζ A

Epist. 1.8.7	sed quia mente *minus validus* quam corpore toto

ζ B

Ars 69	*nedum sermonum* stet honus et gloria vivax

ILIAS LATINA (1070 lines)

	A		B	
α^1		–	1	0.09
α^2	1	0.09	–	
α^5	1	0.09	–	
γ^1	2	0.19	–	
δ^1	5	0.47	4	0.37
δ^2	1	0.09	–	
ε	2	0.19	–	
ζ	2	0.19	–	

The manuscript reading in 1011 *caestibus adversos* (v.l.-*sis*) *cunctos superavit Epeos* is suspicious, since in Homer (*Il.* 23.677) Epeus has only one opponent. Hence Baehrens' suggestion *adversum Euryalum*, which eliminates the α^2. *cunctos* may derive from *omnes* in the next line.

α^1 B

961   *libratum iaculum* Vulcania torquet in arma

α^2 A

1011   caestibus *adversos cunctos* superavit Epeos[16]

α^5 A

679   sic rursus *Danai turbati* caede suorum

γ^1 A

444   cominus hunc *gladio, iaculo* ferit eminus illum
508   funduntur *Teucri, Danai* laetantur ovantes

δ^1 A

88   *pro nato* veni genetrix en ad tua supplex
292   praedonis *Phrygii, ni* vastum ferrea pectus
447   post *hos infestos* Chromiumque et Echemona telo
987   pectora *pro Priamo, pro nostro* corpore Pyrrhus
1044   allevat *a terra* corpusque exsangue parenti

δ^1 B

13   *nam quondam* Chryses, sollemni tempora vitta
826   Dardanides, *quam prolapsam* celere excipit ictu
957   *hastam iam manibus saevus* librabat Achilles
959   *quam praeterlapsam* vitavit callidus Hector

δ^2 A

393   *huc illuc* ensemque ferox hastamque coruscat

ε A

82   *ne se* Plistheniden contra patiatur inultum
573   ut meus hic, *pro quo* tua numina, natus, adoro

ζ A

321   coniugis *antiqui? vidi* puduitque videre
957   *hastam iam manibus saevus* librabat Achilles

---

[16] See above.

JUVENAL (3621 lines)

	A		B	
α^1	3	0.08	2	0.05
α^2	1	0.03	–	
α^5	1	0.03	–	
α^6	1	0.03	–	
γ^1	1	0.03	–	
γ^2	1	0.03	–	
δ^1	32	0.88	7	0.20
ε	5	0.14	–	
ζ	–		2	0.05

Dispondaic endings and endings of one, four, and five syllables are frequent.[17] Five are noun-plus-attribute combinations, three of which are homs, one of them a Greek word. I have omitted all three from the register.[18]

α^1 A

1.68	*exiguis tabulis* et gemma fecerit uda
5.14	fructus *amicitiae magnae* cibus: imputat hunc rex
15.160	*cognatis maculis* similis fera; quando leoni

α^1 B

3.72	viscera *magnarum domuum* dominique futuri
6.O 14	sed tibi *communem calicem* facit uxor et illis

α^2 A

11.19	et *quadringentis nummis* condire gulosum

α^5 A

3.254	scinduntur *tunicae sartae* modo, longa coruscat

α^5 B

10.122	o *fortunatam natam* me consule Romam[19]

---

[17] In the first eight satires I find thirty-seven double spondees and quadrisyllables of which twenty-nine contain names or words of Greek origin. As in other authors, non-hom noun-plus-attribute combinations abound – eighteen in the first hundred lines of Satire I.

[18] See p. 3.

[19] Quoted from Cicero.

α⁶ A

2.119		*signatae tabulae,* dictum 'feliciter', ingens

γ¹ A

2.56		Penelope *melius, levius* torquetis Arachne

γ² A

10.35		*praetextae trabeae* fasces lectica tribunal

δ¹ A

1.55		cum leno accipiat moechi bona, *si capiendi*
1.67		signator *falsi, qui* se lautum atque beatum
1.89		*hos animos?* neque enim loculis comitantibus itur
1.107		optandum, *si Laurenti* custodit in agro
3.19		numen aquis, *viridi si* margine cluderet undas
3.53		carus erit *Verri qui* Verrem tempore quo vult
3.202		*a pluvia,* molles ubi reddunt ova columbae
4.93		solstitia, *his armis* illa quoque tutus in aula
5.109		*a Seneca,* quae Piso bonus, quae Cotta solebat
5.173		flagra pati, *his epulis* et tali dignus amico
6.103		qua tamen exarsit *forma, qua* capta inventa
6.306		i nunc et *dubita qua* sorbeat aera sanna
6.O 13		accipit *has animas* aliosque in carcere nervos
6.376		provocat *a domina* factus spado. dormiat ille
7.122		si quater *egisti, si* contigit aureus unus
8.257		sufficiunt *dis infernis* Terraeque parenti
9.117		*pro populo* faciens quantum Saufeia bibebat
9.131		stantibus et *salvis his* collibus. undique ad illos
10.46		defossa in *loculos. quos* sportula fecit amicos
10.126		volveris *a prima* quae proxima. saevus et illum
10.269		*qui domini* cultris tenue et miserabile collum
10.343		obsequere imperio, *si tanti* vita dierum
10.353		notum *qui pueri* qualisque futura sit uxor
11.88		functus ad *has epulas* solito maturius ibat
11.171		non capit *has nugas* humilis domus. audiat ille
13.170		unguibus *a saeva* fertur grue, si videas hoc
14.14		semper et *a magna* non degenerare culina
14.176		immodici census, nam dives *qui fieri* vult
14.309		non ardent *Cynici; si* fregeris, altera fiet
15.30		nam scelus, *a Pyrrha* quamquam omnia syrmata volvas
15.38		esse *deos quos* ipse colit. sed tempore festo
15.130		supplicia *his populis,* in quorum mente pares sunt

δ^1 B

2.57	autumno, *iam quartanam* sperantibus aegris
3.198	nocte metus. iam poscit *aquam, iam* frivola transfert
6.167	malo *Venusinam quam* te, Cornelia, mater
6.182	usque adeo est, ut non *illam quam* laudibus effert
6.192	tunc *etiam, quam* sextus et octingesimus annus
6.398	sed cantet potius *quam totam* pervolet urbem
11.204	effugiatque *togam.* iam nunc in balnea saeva

ε A

4.70	surgebant cristae. nihil est quod credere *de se*
5.58	et, *ne te* teneam, Romanorum omnia regum
6.566	ante tamen *de te* Tanaquil tua; quando sororem
14.6	nec melius *de se* cuiquam sperare propinquo
16.8	haut minimum illud erit, *ne te* pulsare togatus

ζ B

3.44	nec volo nec *possum; ranarum* viscera numquam
6.014	sed tibi *communem calicem* facit uxor et illis

LAUS PISONIS (261 lines)

	A		B
δ^1	2	0.77	–
ε	1	0.38	–
ζ	1	0.38	–

δ^1 A

154	pocula sumit *ea qua* gessit fulmina dextra
231	qui sonat, *ingenti qui* nomine pulsat Olympum

ε A

250	si tamen hoc ulli *de se* promittere fas est

ζ A

205	interea *sectis quamvis* acerrima surgant

LUCAN (8060 lines)

	A		B	
α^1	1	0.01	–	
α^2	1	0.01	–	
α^4	1	0.01	1	0.01
α^7	2	0.02	3	0.04
β^1	1	0.01	–	
γ^1	2	0.02	1	0.01
γ^2	1	0.01	–	
γ^3	–		1	0.01
δ^1	31	0.46	25	0.31
δ^2	2	0.02	–	
ε	6	0.07	1	0.01
ζ	17	0.21	16	0.20

Avoidance of α^1 is almost total, if not quite; for in 5.109 f. *sustulit iras*/telluris sterilis *monstrato fine* the adjective could be construed with *iras*, 'sterile anger' standing for 'anger causing sterility', much as snakes (9.718, 790) are called parched or thirsty because their victims become so. But the omega count is notably higher than in other silver epicists. On the readings in 4.158 and 821, 6.339, 9.153 see *Proc. Cambr. Phil. Soc.* 33 (1987).89. Unenlightened about homs, Housman argued for and printed the paradosis *festus* in 1.602 *septemvirque epulis festus* and was followed in my edition, but the old vulgate *festis* lingers on.

See also on Valerius Flaccus.

α^1 A

5.110   *telluris sterilis* monstrato fine; resolvit[20]

α^2 A

2.194   Praenestina *suos cunctos* simul ense recepto

α^4 A

5.43   det vobis *animos, quantos* fugientibus hostem

α^4 B

7.456   *vindictam, quantam* terris dare numina fas est

[20] See above.

$$\alpha^7 A$$

3.275	*Europae, mediae* dirimens confinia terrae
4.244	*invidia caeca* bellorum in nocte tulisset

$$\alpha^7 B$$

5.648	non valet in *fluctum: victum* latus unda repellens
6.53	sufficit in *regnum, subitum* bellique tumultu
8.718	ausus ferre *gradum victum* pietate timorem

$$\beta^1 A$$

8.839	ossa *tui Magni, si* nondum subruta fluctu

$$\gamma^1 A$$

10.59	dedecus *Aegypti, Latii* feralis Erinys
10.291	nunc Arabum *populis, Libycis* nunc aequus harenis

$$\gamma^1 B$$

5.71	Hesperio *tantum, quantum* summotus Eoo

$$\gamma^2 A$$

4.806	ferre *datis, luitis* iugulo sic arma, potentes

$$\gamma^3 B$$

7.114	pugnatur. *quantum scelerum* quantumque malorum

$$\delta^1 A$$

1.284	pars quota terrarum: *facili si* proelia pauca
1.299	bellorum o *socii, qui* mille pericula Martis
1.381	castra super *Tusci si* ponere Thybridis undas
1.463	et *vos, crinigeros* Belgis arcere Caycos
2.555	te quoque *si superi* titulis accedere nostris
2.561	dux sit in *his castris* senior, dum miles in illis
3.191	*Penei qui* rura colunt, quorumque labore
3.366	armorum, nisi *qui vinci* potuere rebellant
3.430	maiestate *loci, si* robora sacra ferirent
4.116	laxet iter fluviis: *hos campos* Rhenus inundet
4.382	heu *miseri, qui* bella gerunt! tunc arma relinquens
4.409	cautus ab incursu *belli, si* sola recedat
5.219	dumque a luce *sacra, qua* vidit fata, refertur
5.273	paupertate pii. *finis quis* quaeritur armis
5.321	hic fuge, *si belli* finis placet, ense relicto

5.336	damnum posse fugae? *veluti, si* cuncta minentur
5.362	et tu, *quo solo* stabunt iam robore castra
5.576	prendere *ne longe* nimium sit proximtellus
6.70	ipse quoque *a tuta* deducens agmina Petra
6.611	at simul *a prima* descendit origine mundi
7.185	quid mirum *populos quos* lux extrema manebat
7.400	humani! toto *populi qui* nascimur orbe
7.560	inspicit et gladios, *qui toti* sanguine manent
7.579	scit cruor *imperii qui* sit, quae viscera rerum
7.641	vincitur *his gladiis* omnis quae serviet aetas
7.806	aut, *generi si* poena iuvat, nemus extrue Pindi
8.94	*a causa* meliore deos. o maxime coniunx
8.297	segnis, et *a nulla* mors est incerta sagitta
8.328	consilium damnasse *viros; quos* Lentulus omnis
8.389	credis, Magne, *viros, quos* in discrimina belli
8.392	*a terra* moriare tua, tibi barbara tellus
8.544	*hos animos?* sic fata premunt civilia mundum
8.626	ignorant *populi, si* non in morte probaris
8.781	quam metuis, demens, *isto pro* crimine poenam
8.839	ossa *tui Magni, si* nondum subruta fluctu
9.682	Perseos *aversi; si* non Tritonia densos
9.972	unde puer raptus *caelo, quo* vertice Nais

δ¹B

1.275	traximus *imperium, tum* cum mihi rostra tenere
1.328	quas nemore Hyrcano *matrum dum* lustra secuntur
1.561	*tum pecudum* faciles humana ad murmura linguae
2.60	*dum nondum* meruere, feri. tantone novorum
2.433	*Hesperiam, quam* cum Scyllaeis clauditur undis
3.168	pauperiorque fuit *tum primum* Caesare Roma
3.443	*tum primum* posuere comas et fronde carentes
5.514	et *iam turrigeram* Bruti comitata carinam
3.647	*dum nimium* pugnax unius turba carinae
4.692	*dum regnum* te, Roma, facit. memor ille doloris
4.742	fraude sua cessere *parum, dum* colle relicto
5.190	spumea *tum primum* rabies vesana per ora
5.632	*tum superum* convexa tremunt atque arduus axis
5.806	insomnis; viduo *tum primum* frigida lecto
6.614	stat genus *humanum, tum,* Thessala turba fatemur
6.667	pectora *tum primum* ferventi sanguine supplet
7.677	parte absente mori. *tum Magnum* concitus aufert
8.66	semianimem conantur *eram; quam* pectore Magnus

8.151   moverat ast *illam, quam* toto tempore belli
9.51   *nam, postquam* frustra precibus Cornelia nautas
9.276   *quam vitam* veniamque libet? rapiatur in undas
9.282   quin agite et magna *meritum cum* caede parate
9.324   *tum, quarum* recto deprendit carbasa malo
9.417   orbis abit *Asiam. nam* cum communiter istae
9.605   et plaga, *quam nullam* superi mortalibus ultra

δ^2 A

2.619   *hinc illinc* montes scopulosae rupis aperto
8.699   *huc illuc* iactatur aquis. adeone molesta

ε A

3.276   *nunc hunc, nunc* illum, qua flectitur, ampliat orbem
6.787   et Curios, Sullam *de te*, Fortuna, quarentem
7.34   illa rati semper *de te* sibi conscia voti
7.83   expectat: propera, *te ne* tua classica linquant
9.854   quod peream, caeloque mori. nil, Africa, *de te*
9.855   nec *de te*, natura, queror: tot monstra ferentem

ε B

9.317   tellus Syrtis erit; *nam iam* brevis unda superne

ζ A

1.407   ius habet aut *Zephyrus, solus* sua litora turbat
2.19   omnis *honos, nullos* comitata est purpura fasces
2.169   meque ipsum *memini caesi* deformia fratris
2.311   excipiam *medius totius* vulnera belli
4.26   *exposuit. piguit* sceleris; pudor arma furentum
4.476   libera non *ultra parva* quam nocte iuventus
4.634   nec sic *Inachiis, quamvis* rudis esset, in undis
4.659   haec fuit. en, *veteris cernis* fastigia valli
4.760   nec *quamvis crebris* iussi calcaribus addunt
5.290   dux erat, hic *socius; facinus* quos inquinat aequat
5.559   iussa, vel hoc *potius pelagus* flatusque negabunt
7.71   *affusi, vinci* socerum patiare rogamus
7.110   res mihi *Romanas dederas*, Fortuna, regendas
7.716   pandunt templa, *domos, socios* se cladibus optant
8.464   infimaque *Aegypti pugnaci* litora velo
8.538   permittant *famuli, sceleri* delectus Achillas
10.274   Aethiopum *lectos: illos* rubicunda perusti

ζB

1.543	desperare *diem; qualem* fugiente per ortus
2.298	natorum *orbatum longum* producere funus
3.74	*Gallorum tantum* populis Arctoque subacta
3.744	*confodiam. veniam* misero concede parenti
4.681	*Medorum, tremulum* cum torsit missile, Mazax
5.15	ut *primum maestum* tenuere silentia coetum
5.648	non valet in *fluctum: victum* latus unda repellens
7.566	vulnera *multorum totum* fusura cruorem
7.861	navita, nec *terram quisquam* movisset arator
8.106	coniugis in *gremium cunctorum* lumina solvit
8.257	et sibi *consultum? Cilicum* per litora totus
8.431	non tibi, cum *primum gelidum* transibis Araxin
9.134	nec credens *Pharium tantum* potuisse tyrannum
9.532	solstitii *medium signorum* percutit orbem
9.802	*intactum volucrum* rostris epulasque daturum
9.1075	non tuleram *Magnum mecum* Romana regentem

LUCRETIUS (7415 lines)

	A		B	
α^1	31	0.42	18	0.24
α^2	9	0.12	12	0.16
α^3	7	0.09	–	
α^4	2	0.03	1	0.01
α^5	3	0.04	1	0.01
α^7	3	0.04	2	0.03
β^1	5	0.07	1	0.01
β^2	8	0.11	–	
γ^1	3	0.04	–	
γ^2	3	0.04	–	
γ^3	2	0.03	3	0.04
δ^1	28	0.38	23	0.31
δ^2	1	0.01	–	
ε	6	0.08	1	0.01
ζ	14	0.19	17	0.23

Lucretius being given to formulae, I have counted the following only once:
quo pacto, concursus motus, nequiquam quoniam, veluti pueri, a vera ... ratione.
Reckoned thus, alpha and omega counts are almost equal.

α^1 A

1.138	multa *novis verbis* praesertim cum sit agendum
1.771	corpus et *aerias auras* roremque liquoris
1.920	et *lacrimis salsis* umectent ora genasque
2.26	lumina *nocturnis epulis* ut suppeditentur
2.42	*subsidiis magnis* et ecum vi constabilitas
2.98	partim *intervallis magnis* convecta resultant
2.107	in *magnis intervallis;* haec aera rerum
2.552	sed quasi *naufragiis magnis* multisque coortis
2.622	*ingratos animos* atque impia pectora vulgi
2.816	sed *variis formis* variantis edere tactus
2.1054	seminaque *innumero numero* summaque profunda
3.1	e *tenebris tantis* tam clarum extollere lumen
3.509	cum *validis ventis* aetatem degere posse
3.568	nec *magnis intervallis* primordia possunt
3.628	possumus *infernas animas* Acherunte vagare
3.717	sin ita *sinceris animis* ablata profugit
3.779	*innumero numero* certareque praeproperanter
4.609	et simulacra *viis derectis* omnia tendunt
4.1285	vincitur in *longo spatio* tamen atque labascit
5.345	nam cum res *tantis morbis* tantisque periclis
5.528	in *variis mundis* varia ratione creatis
5.583	praebet, ut est *oris extremis* cumque notata
5.695	omnia *dispositis signis* ornata notarunt
5.880	ex *alienigenis membris* compacta, potestas
6.124	cum subito *validi venti* collecta procella
6.163	exsilit et *claras scintillas* dissipat ignis
6.231	curat item *vasis integris* vina repente
6.363	tum *variae causae* concurrunt fulminis omnes
6.622	largiter in *tanto spatio* tamen auferet undis
6.1011	quam *validi ferri* natura et frigidus horror
6.1256	*exanimis pueris* super exanimata parentum

α^1 B

1.383	unde *initum primum* capiat res quaeque movendi
1.523	omne quod est *spatium vacuum* constaret inane
1.639	clarus ob *obscuram linguam* magis inter inanis
1.779	*naturam clandestinam* caecamque adhibere
2.123	dumtaxat *rerum magnarum* parva potest res
2.358	conspicere *amissum fetum,* completque querellis
2.379	unius ad *certam formam* primordia rerum

2.1062	*magnarum rerum* fierent exordia semper
2.1132	frangit et in *partem peiorem* liquitur aetas
3.649	nec tenet *amissam laevam* cum tegmine saepe
3.855	*praeteritum spatium, tum* motus materiai
3.1003	deinde animi *ingratam naturam* pascere semper
3.1047	qui somno *partem maiorem* conteris aevi
5.291	et *primum iactum* fulgoris quemque perire
5.430	*magnarum rerum* fiunt exordia saepe
6.26	exposuitque *bonum summum* quo tendimus omnes
6.131	saepe ita dat *parvum sonitum* displosa repente
6.135	scilicet ut, *crebram silvam* cum flamina cauri

$\alpha^2 A$

1.341	multa *modis multis* varia ratione moveri
2.720	nam veluti *tota natura* dissimiles sunt
2.778	ut saepe ex *aliis formis* variisque figuris
3.110	non *alio pacto* quam si, pes cum dolet aegri
3.340	qui datus est, neque *ea causa* convellitur ipse
3.1027	inde *alii multi* reges rerumque potentes
5.1070	longe *alio pacto* gannitu vocis adulant
6.892	et *multis aliis* praebet regionibus aequor
6.1162	dissoluebat *eos, defessos* ante fatigans

$\alpha^2 B$

2.696	*multarum rerum* cum sint primordia, verum
2.846	nec iaciunt *ullum proprium* de corpore odorem
2.938	sensus ante *ipsam genitam* naturam animantis
3.137	inter se atque *unam naturam* conficere ex se
3.153	consentire *animam totam* per membra videmus
3.216	ergo *animam totam* perparvis esse necesse est
3.268	omnibus est *unum perfectum* corporis augmen
3.704	disperit atque *aliam naturam* sufficit ex se
4.149	scinditur ut *nullum simulacrum* reddere possit
4.352	possint *ullarum rerum* coniecta movere
5.536	convenit atque *aliam naturam* subter habere
5.1367	inde aliam atque *aliam culturam* dulcis agelli

$\alpha^3 A$

1.800	posse eadem *demptis paucis* paucisque tributis
2.275	perspicuum est *nobis invitis* ire rapique
3.171	ossibus ac *nervis disclusis* intus adacta
3.368	iam magis *exemptis oculis* debere videtur

4.437	navigia *aplustris fractis* obnitier undis
4.761	*relicta vita* iam mors et terra potita est
4.1105	denique cum *membris collatis* flore fruuntur

$$\alpha^4\,A$$

| 2.223 | nec foret *offensus natus* nec plaga creata |
| 5.519 | sive quod *inclusi rapidi* sunt aetheris aestus |

$$\alpha^4\,B$$

| 4.557 | at si *interpositum spatium* sit longius aequo |

$$\alpha^5\,A$$

5.487	tam magis *expressus salsus* de corpore sudor
6.186	*exstructis aliis* alias super impete miro
6.1162	dissoluebat *eos, defessos* ante fatigans

$$\alpha^5\,B$$

| 5.1145 | nam genus *humanum defessum* vi colere aevum |

$$\alpha^7\,A$$

2.743	ex ineunte aevo *nullo coniuncta* calore
5.721	dimidia ex *parti candenti* lumine tinctus
6.1233	deficiens *animo maesto* cum corde iacebat

$$\alpha^7\,B$$

| 2.851 | *naturam, nullam* quae mittat naribus auram |
| 6.1247 | inque aliis *alium, populum* sepelire suorum |

$$\beta^1\,A$$

2.102	indupedita *suis perplexis* ipsa figuris
4.1019	indicioque *sui facti* persaepe fuere
4.1274	idque *sua causa* consuerunt scorta moveri
6.618	exsiccare *suis radiis* ardentibu' solem
6.1283	namque *suos consanguineos* aliena rogorum

$$\beta^1\,B$$

| 6.983 | esse et habere *suam naturam* quaeque viasque |

$$\beta^2\,A$$

1.856	ex *oculis nostris* aliqua vi victa perire
3.276	quod genus in *nostris membris* et corpore toto
4.252	et *nostros oculos* perterget longior aura
4.285	pervenit et *nostros oculos* reiecta revisit

4.759	haec eadem *nostros animos* quae cum vigilamus
4.881	dico *animo nostro* primum simulacra meandi
5.578	quam, *nostris oculis* qua cernimus, esse videntur
5.874	*praesidio nostro* pasci genus esseque tutum

$$\gamma^1\,\mathrm{A}$$

1.659	ardua dum *metuunt, amittunt* vera viai
2.73	unde *abeunt minuunt,* quo venere augmine donant
4.1267	pectoribus *positis, sublatis* semina lumbis

$$\gamma^2\,\mathrm{A}$$

1.634	*concursus motus,* per quae res quaeque geruntur
2.905	visceribus *nervis venis,* quaecumque videmus
5.1353	insilia ac *fusi radii* scapique sonantes

$$\gamma^3\,\mathrm{A}$$

3.487	quin etiam subito *vi morbi* saepe coactus
3.492	nimirum quia *vi morbi* distracta per artus

$$\gamma^3\,\mathrm{B}$$

3.198	at contra *lapidum coniectum* spicarumque
3.901	iam *desiderium rerum* super insidet una
5.150	quae quoniam *manuum tactum* suffugit et ictum

$$\delta^1\,\mathrm{A}$$

1.84	Aulide *quo pacto* Triviai virginis aram
1.105	somnia *quae vitae* rationes vertere possint
1.648	nec *rarefieri, si* partes ignis eandem
1.785	ex imbri retroque *a terra* cuncta reverti
2.82	avius *a vera* longe ratione vagaris
2.198	derecta et magna *vi multi* pressimus aegre
2.1046	quid sit ibi *porro quo* prospicere usque velit mens
3.296	quo genere in *primis vis* est violenta leonum
3.394	et quam in *his intervallis* tuditantia possint
3.733	corpus enim magis *his vitiis* affine laborat
3.867	nec miserum *fieri qui* non est posse neque hilum
3.989	obtineat, sed *qui terrai* totius orbem
3.1025	lumina *sis oculis* etiam bonus Ancu' reliquit
4.413	*quae variae* retinent gentes et saecla ferarum
4.935	proptereaque fere *res omnes* aut corio sunt
4.1009	accipitres somno in *leni si* proelia pugnas
4.1172	cui Veneris *membris vis* omnibus exoriatur
5.350	atque illi quos *a vita* natura removit

5.510	principio magnus *caeli si* vertitur orbis
5.544	pondera *sunt, laedunt,* permulto saepe minora
5.904	*qui fieri* potuit, triplici cum corpore ut una
5.1197	vulnera, *quas lacrimas* peperere minoribu' nostris
6.136	*perflant, dant* sonitum frondes ramique fragorem
6.846	et quasi *concrescit, fit* scilicet ut coeundo
6.953	ignis, *qui ferri* quoque vim penetrare suevit
6.1033	continuo fit *uti qui* post est cumque locatus
6.1034	denique *res omnes* debent in corpore habere
6.1075	corpore cum lanae, *dirimi qui* non queat usquam

δ^1 B

1.393	errat; nam *vacuum tum* fit quod non fuit ante
1.448	nec ratione animi *quam quisquam* possit apisci
1.584	denique *iam quoniam* generatim reddita finis
1.649	*naturam quam* totus habet super ignis haberent
1.698	quod mihi cum *vanum tum* delirum esse videtur
2.46	*tum vacuum* pectus linquunt curaque solutum
2.83	*nam quoniam* per inane vagantur, cuncta necesse est
2.265	vim *cupidam tam* de subito quam mens avet ipsa
2.536	sicut *quadrupedum cum* primis esse videmus
3.42	infamemque ferunt *vitam quam* Tartara leti
3.57	nam verae voces *tum demum* pectore ab imo
3.397	et dominantior ad *vitam quam* vis animai
3.427	principiis *factam quam* liquidus umor aquai
3.855	*praeteritum spatium, tum* motus materiai
3.1085	posteraque in dubio est *fortunam quam* vehat aetas
4.884	rem *quisquam, ⟨quam⟩* mens providit quod velit ante
4.919	dissoluuntur enim *tum demum* membra fluuntque
4.968	nautae *contractum cum* ventis degere bellum
5.602	nonne vides *etiam quam* late parvus aquai
5.888	*tum demum* puerili aevo florente iuventas
5.1169	quippe etenim iam *tum divum* mortalia saecla
6.282	impetus incessit, *maturum tum* quasi fulmen
6.936	nunc omnes *repetam quam* raro corpore sint res

δ^2 A

5.881	*hinc illinc* partis ut sat par esse potissit

ε A

1.370	illud in his rebus *ne te* deducere vero
2.1124	plura sibi assumunt quam *de se* corpora mittunt

4.8	deinde quod obscura *de re* tam lucida pango
5.110	qua prius aggrediar quam *de re* fundere fata
6.768	percipe; nam *de re* nunc ipsa dicere conor
6.979	hoc etiam superest, ipsa quam dicere *de re*

ε B

6.1080	cetera *iam quam* multa licet reperire? quid ergo?

ζ A

1.539	sunt ita *uti docui*, sint haec aeterna necesse est
2.55	nam *veluti pueri* trepidant atque omnia caecis
2.137	ipsaque *proporro paulo* maiora lacessunt
2.327	excitur *pedibus sonitus* clamoreque mentes
2.562	disiectare aestus *diversi materiai*
3.182	nil adeo *fieri celeri* ratione videtur
3.620	atque ita *multimodis partitis* artubus esse
4.212	ponitur, *extemplo caelo* stellante serena
4.773	scilicet id *fieri celeri* ratione putandum est
6.43	et quoniam *docui mundi* mortalia templa
6.147	ignem, *continuo magno* clamore trucidat
6.423	quod *superest, facile est* ex his cognoscere rebus
6.534	cum bene *cognoris elementis* reddita quae sunt
6.1218	aut ubi *gustarat, languebat* morte propinqua

ζ B

1.668	occidet ad *nilum nimirum* funditus ardor
1.723	murmura *flammarum rursum* se colligere iras
1.956	pervideamus *utrum finitum* funditus omne
1.971	id validis *utrum contortum* viribus ire
1.1047	principiis *rerum spatium* tempusque fugai
1.1089	atque ideo *totum circum* tremere aethera signis
2.217	corpora cum *deorsum rectum* per inane feruntur
2.635	cum pueri *circum puerum* pernice chorea
2.808	qui *quoniam quodam* gignuntur luminis ictu
2.1053	undique cum *versum spatium* vacet infinitum
2.1148	*nequiquam, quoniam* nec venae perpetiuntur
3.208	haec quoque res *etiam naturam* dedicat eius
3.579	atque *animam, quoniam* coniuncta est causa duobus
4.361	fit quasi ut ad *tornum saxorum* structa tuantur
5.1014	tum genus *humanum primum* mollescere coepit
5.1362	ipsa fuit *rerum primum* natura creatrix
6.1093	expediam. *primum multarum* semina rerum

LYGDAMUS (290 lines)

	A		B
γ^1	1	0.37	–
ζ	1	0.37	–

290 lines is not much to go by, but perhaps enough to discountenance *proles Semeles* in 4.45, where *Semelae* is a variant. In Tib. 1.2.54 Luck reads *Hecatae,* referring to Housman on Manil. 4.469 (corrected reference). Housman's note shows that after Virgil and Horace, who always use the Latin forms, *–es* secured by metre (i.e. before a word beginning with a vowel) occurs sixty times, *–ae* elided before a vowel five times, including three of *Libyae* in Silius. But that does not tell us what happened before words beginning with a consonant.

$$\gamma^1\, A$$

5.31   vivite *felices, memores* et vivite nostri

$$[\gamma^3\, A$$

4.45   sed *proles Semeles* Bacchus doctaeque sorores]

$$\zeta\, A$$

5.7   non ego *temptavi nulli* temeranda virorum

MANILIUS (4258 lines)

	A		B	
α^1	19	0.44	5	0.12
α^2	7	0.16	1	0.02
α^3	3	0.07	–	
α^4	–		1	0.02
α^5	1	0.02	–	
α^6	1	0.02	–	
β^1	2	0.05	1	0.02
β^2	1	0.02	–	
γ^3	–		2	0.05
δ^1	20	0.47	11	0.26
δ^2	1	0.02	–	
ζ	9	0.21	6	0.14

Although the nature of his subject might be thought to have something to do with his high tolerance of alphas, this extends, as with Lucretius, to non-technical passages.

α¹ A

1.699	trisque secat *medios gyros* et signa ferentem
2.67	quod nisi *cognatis membris* contexta maneret
2.71	erraretque *vagus mundus* standove rigeret
2.185	ille *senescentis veris,* subeuntis at ille
2.454	singulaque *im⟨periis⟩ propriis* parentia membra
2.578	quae sunt *adversis signis* cognata trigona
2.898	quaeritur *inversu⟨s⟩ titulus.* sub corde sagaci
3.456	ac *medii numeri* coniuncta sequentibus astris
4.187	ingenium ad *subitas iras* facilisque recessus
4.293	bis sex *materia propria* pollentia signa
4.581	anguipedem *alatos umeros* Typhona ferentem
4.907	*sidereos oculos* propiusque aspexit Olympum
5.220	*effrenos animos* violentaque pectora fingit
5.368	ipse quoque *aerios populos* caeloque dicatum
5.487	*rorantis iuvenis,* quem terris sustulit ipsa
5.604	*sanguineis undis* pontumque exstillat in astra
5.617	mercedem *tanti belli,* quo concidit ipsa
5.643	quadrupedum, et *medicas herbas* in membra ferarum
5.722	cumque *vagae stellae*[21] terris sua lumina condent

α¹ B

1.767	Auroraeque *nigrum partum,* stirpemque Tonantis
2.183	sufficit, *aestatem sitientem* provehit alter
4.859	bina per *adversum caelum* fulgentia signa
5.71	ille dabit *proprium studium* caeloque retentas
5.289	condentemque *novum caelum* per tecta Tonantis

α² A

2.299	aut tribus ac *binis signis* ornare trigonem
3.183	at si *subiectis senis* fulgebit in astris
3.567	bis *quinos annos* Aries unumque triente
3.602	*tricenos annos* duplicat, tris insuper addit
4.410	per *denos numeros* et quae sint insita cuique
4.488	quaeve auget *quinto numero* vel septima fertur
4.496	et quinta in *quinos numeros* revocata duasque

α² B

1.900	infecitque *trium legionum* sanguine campos

[21] Conjectural.

$$\alpha^3\,A$$

2.674	quae tribus *emensis signis* facit astra trigona
3.463	et tribus *expletis horis* noctemque diemque
4.475	quartaque bis *denis actis* et septima et ambae

$$\alpha^4\,B$$

1.646	nos primam ac *summam sextam* numeramus utramque

$$\alpha^5\,A$$

4.161	in tot *fecundi Gemini* commenta feruntur

$$\alpha^6\,A$$

1.509	quot *capti populi!* quotiens Fortuna per orbem

$$\beta^1\,A$$

1.109	attribuitque *suas formas,* sua nomina signis
2.614	expositumque *suae noxae,* nec fraudibus ullis

$$\beta^1\,B$$

3.496	quave in parte *suam summam* nomenque relinquet

$$\beta^2\,A$$

4.920	ipse vocat *nostros animos* ad sidera mundus

$$\gamma^3\,B$$

2.460	sub Cancro est, *laterum regnum* scapulaeque Leonis
4.629	atque his *undarum tractum* constringit harenis

$$\delta^1\,A$$

1.486	ut voluit *credi, qui* primus moenia mundi
1.604	inter se *adversi, qui* cunctos ante relatos
2.126	et *privas quas* dant leges nascentibus astra
2.220	Lanigeri, sex *a Libra* nocturna videri
2.256	quod *si sollerti* circumspicis omnia cura
2.321	quod, cum totius *numeri, qui* construit orbem
2.333	*quadrati si* forte voles effingere formam
2.433	*his animadversis* rebus quae proxima cura
2.717	cetera *pro numero* ducunt ex ordine partes
2.729	inde suas illi *signo, quo* Luna refulsit
2.810	primus erit, *summi qui* regnat culmine caeli
2.950	tergaque prospectat *Phoebi, qui* viderat ora
2.951	ne mirere, *nigri si* Ditis ianua fertur

3.56	exceptum *a summa* ne quid ratione maneret
3.221	ut parte ex *illa qua* Phoebi coeperit orbis
3.341	*pro spatio* mora magna datur: quae proxima nobis
3.465	incipit *a sexta* tempus procedere parte
4.393	*pro pretio* labor est nec sunt immunia tanta
5.358	Arcturum ostendit *ponto. quo* tempore natis
5.743	cui si *pro numero* vires natura dedisset

δ¹ B

1.620	quam septem stellae *primam iam* sole remoto
2.364	a Tauro venit in *Cancrum, tum* Virgine tacta
2.782	*tum demum* consurgit opus, cum cuncta supersunt
3.449	*quam minimam* Capricornus agit, noctisque per horas
3.450	*quam summam;* quodque a iusto superaverit umbris
3.652	*tum primum* miti pelagus consternitur unda
3.654	*tum pecudum* volucrumque genus per pabula laeta
4.648	pervenit et patulis *tum demum* funditur arvis
4.851	sicut Luna suo *tum tantum* deficit orbe
5.104	hirtaque *tum demum* terris promittere terga
5.716	*tum quartum* sextumque genus discernitur omni

δ² A

1.199	*huc illuc* agilis, et servet in aethere metas

ζ A

2.368	*alterius ductus* locus est per transita signa
2.531	consilium est *maius. victus* Leo fulget in astris
2.642	in terris *geniti tali* sub lege creantur
3.66	undique *uti fati* ratio traheretur in unum
3.237	omnia signa *pari mundi* sub lege meare
3.299	partibus ut *ratio signo* ducatur ab illo
4.196	*quamvis occultis* naturae condita causis
4.213	non alio *potius genitus* sit Servius astro
5.214	atque eget *alterius mundus;* natura suismet

ζ B

1.84	in commune *bonum commentum* laeta dederunt
2.764	effluat in *vanum rerum* praeposterus ordo
2.785	materies *primum rerum,* ratione remota
3.586	sed mihi *templorum tantum* nunc iura canentur
4.274	his erit in *pontum studium,* vitamque profundo
5.356	et sibi non *aegrum iamdudum* credere corpus

MARTIAL (6657 lines)

	A		B	
α^1	2	0.03	–	
γ^1	4	0.06	1	0.02
γ^2	3	0.04	1	0.02
γ^3	–		5	0.08
δ^1	61	0.92	19	0.30
δ^2	3	0.04	--	
ε	8	0.12	--	
ζ	5	0.08	11	0.17

A strict avoider of alphas, with only two in 6657 lines, both α^1s and both in the same eight-line epigram. His omega count, however, is rather high for a classical poet.

In Sp. 9(7). 7 the standard supplement, *denique supplicium* ⟨*dignum tulit: ille parentis*⟩ (Scheidewin) no longer stands: substitute *dignus* or *merito*.

α^1 A

2.7.2	*historias bellas,* carmina bella facis
2.7.4	*bellus grammaticus,* bellus es astrologus

γ^1 A

5.21.1	Quintum *pro Decimo, pro Crasso,* Regule, Macrum
5.43.1	Thais habet *nigros, niveos* Laecania dentes
10.89.3	ore nitet *tanto, quanto* superasset in Ide
12.46(47).1	*difficilis, facilis,* iucundus acerbus es idem

γ^1 B

6.51.4	'quid facies?' inquis. quid *faciam? veniam*

γ^2 A

1.68.3	*cenat, propinat,* poscit, negat, innuit: una est
6.60(61).1	laudat, *amat, cantat* nostros mea Roma libellos
7.61.9	tonsor, copo, *cocus, lanius* sua limina servant

γ^2 B

12.72.5	*frumentum, milium* tisanamque fabamque solebas

γ^3 B

1.4.2	*terrarum dominum* pone supercilium
1.60.5	quid frustra *nemorum dominum* regemque fatigas

1.76.1	o mihi *curarum pretium* non vile mearum
4.32.3	*dignum tantorum pretium* tulit illa laborum
7.5.5	*terrarum dominum* propius videt ille tuoque

δ¹ A

Sp. 17(15).4	primus in *Arctoi qui* fuit arce poli
1.8.6	hunc volo, *laudari qui* sine morte potest
1.19.1	*si memini,* fuerant tibi quattuor, Aelia, dentes
1.58.1	milia *pro puero* centum me mango poposcit
1.70.11	protinus *a laeva* clari tibi fronte Penates
2.53.2	sed *fieri si* vis, hac ratione potes
2.63.3	Miliche, luxuria est *si tanti* dives amares
2.64.6	si quid habes *animi, si* quid in arte vales
2.66.3	hoc facinus Lalage *speculo, quo* viderat, ulta est
2.75.10	*a nostra* pueris parcere disce lupa
2.77.1	*Cosconi, qui* longa putas epigrammata nostra
2.91.7	haec, *si displicui,* fuerint solacia nobis
2.91.8	haec fuerint nobis praemia, *si placui*
2.93.3	tu tamen hunc *fieri si* mavis, Regule, primum
3.63.5	cantica *qui Nili, qui* Gaditana susurrat
3.63.10	pallia *vicini qui* refugit cubiti
3.91.12	at nunc *pro cervo* mentula supposita est
4.1.10	*pro tanto* quae sunt improba vota deo
4.18.3	in iugulum *pueri, qui* roscida tecta subibat
4.45.1	haec tibi *pro nato* plena dat laetus acerra
4.88.1	nulla remisisti *parvo pro* munere dona
5.3.2	a famulis *Histri qui* tibi venit aquis
5.16.5	nam *si falciferi* defendere templa Tonantis
5.21.1	Quintum *pro Decimo, pro Crasso,* Regule, Macrum
5.48.8	tardaque *pro tanto* munere, barba, veni
6.64.6	emendare *meos, quos* novit fama, libellos
6.87.2	di mihi dent et tu quae volo, *si merui*
7.12.8	*qui Phoebi* radios ferre diemque negat
7.51.11	hunc licet *a decima*-neque enim satis ante vacabit-
7.92.2	uno *bis dicis,* Baccara, terque die
8.70.8	carmina *qui docti* nota Neronis habet
8.75.2	Lingonus *a Tecta* Flaminiaque recens
9.4.2	et plus quam *futui, si* totidem addideris
9.8.2	annua, *si memini,* milia sena dabas
9.21.2	agrum *pro puero* Calliodorus habet
9.31.9	quae litat *argento pro* te, non sanguine, Caesar
9.59.18	quidquid et *a nivea* grandius aure sonat

9.84.2	staret *pro domino* Caesare sancta fides
9.92.5	Gaius *a prima* tremebundus luce salutat
9.101.8	abstulit, *a Stygia* cum cane venit aqua
10.15(14).9	nil aliud *video quo* te credamus amicum
10.28.7	at tu, sancte pater, *tanto pro* munere gratus
10.33.5	ut tu, *si viridi* tinctos aerugine versus
10.41.2	deseris atque *iubes res* sibi habere suas
10.48.8	attulit et *varias quas* habet hortus opes
10.82.2	mane vel *a media* nocte togatus ero
10.95.2	hi, puto, non *dubie se* futuisse negant
11.21.4	quam vetus *a crassa* calceus udus aqua
11.23.15	si potes ista *pati, si* nil perferre recusas
11.42.2	lemmata. *qui fieri,* Caeciliane, potest
11.48.2	iugera *facundi qui* Ciceronis habet
12.2(3).1	ad populos *mitti qui* nuper ab urbe solebas
12.9.3	ergo agimus laeti *tanto pro* munere grates
12.29(26).3	quod non *a prima* discurram luce per urbem
12.38.5	*uxori qui* saepe tuae comes improbus haeret
12.44.7	nec deerant *Zephyri, si* te dare vela iuvaret
12.66.8	stant pueri *dominos quos* precor esse meos
13.103.2	essem *si scombri,* non tibi missa forem
14.1.10	quos tibi *pro caelo* filius ipse dedit
14.65.2	sumere, *pro puero* pes erit ipse tibi
14.160.2	haec *pro Leuconico* stramina pauper emit

$\delta^1 B$

Sp. 3.11	vox diversa sonat *populorum, tum* tamen una est
1.103.3	qualiter o *vivam, quam* large quamque beate
2.18.2	tu capias *aliam: iam* sumus ergo pares
2.60.2	*supplicium tantum dum* puerile times
2.64.1	dum modo *causidicum, dum* te modo rhetora fingis
4.11.1	*dum nimium* vano tumefactus nomine gaudes
4.72.4	'non' inquis *'faciam tam* fatue.' nec ego
5.50.7	desine *iam nostram,* precor, observare culinam
7.46.1	commendare *tuum dum* vis mihi carmine munus
7.77.2	non *faciam: nam* vis vendere, non legere
7.82.3	hunc ego *credideram – nam* saepe lavamur in unum
7.88.10	decipimur: *credam iam,* puto, Lause, tibi
8.11.1	pervenisse *tuam iam* te scit Rhenus in urbem
8.37.2	milia te *centum num* tribuisse putas
9.28.10	Roma sui *famulum dum* sciat esse Iovis
11.41.1	indulget pecori *nimium dum* pastor Amyntas

13.1.6	senio nec *nostrum cum* cane quasset ebur
13.72.1	Argoa *primum sum* transportata carina
14.69(68).2	clara Rhodos *coptam quam* tibi misit edat

δ^2 A

3.63.9	qui legit *hinc illinc* missas scribatque tabellas
9.29.10	quae sciet *hos illos* vendere lena toros
12.60.11	excipere *hos illos* et tota surgere cena

ε A

Sp. 27(24).3	*ne te* decipiat ratibus navalis Enyo
1.58.3	hoc dolet et queritur *de me* mea mentula secum
7.35.3	sed meus, ut *de me* taceam, Laecania, servus
7.54.1	semper mane mihi *de me* mera somnia narras
8.31.1	nescio quid *de te* non belle, Dento, fateris
9.2.2	et queritur *de te* mentula sola nihil
9.59.4	non *hos quos* primae prostituere casae
11.102.7	audiat aedilis *ne te* videatque caveto

ζ A

Sp. 2.1	hic ubi *sidereus propius* videt astra colossus
3.27.1	numquam me *revocas, venias* cum saepe vocatus
5.10.5	sic veterem *ingrati Pompei* quaerimus umbram
6.10.2	'ille *dabit*' dixit 'qui mihi templa dedit'
14.140(139).2	*indueras albas,* exue callainas

ζ B

1.32.2	hoc *tantum possum* dicere, non amo te
2.36.4	nolo *virum nimium,* Pannyche, nolo parum
2.60.2	*supplicium tantum dum* puerile times
3.91.11	*suppositam quondam* fama est pro virgine cervam
4.3.1	aspice quam *densum tacitarum* vellus aquarum
4.32.3	*dignum tantorum pretium* tulit illa laborum
11.39.15	desine, non *possum libertum* ferre Catonem
11.105.2	*saltem semissem,* Garrice, solve mihi
12.46(47).2	nec *tecum possum* vivere nec sine te
13.23.2	ipse *merum secum* portat et ipsa salem
14.187.1	hoc *primum iuvenum* lascivos lusit amores

OVID

Amores (2460 lines)

	A		B	
α^1	1	0.04	–	
β^1	4	0.17	–	
β^2	1	0.04	–	
γ^2	1	0.04	–	
δ^1	21	0.85	6	0.24
δ^2	1	0.04	–	
ε	7	0.28	–	
ζ	6	0.24	5	0.20

Ars, Remedia Amoris De medicamine facie (3244 lines)

	A		B	
α^2	1	0.03	–	
β^1	2	0.06	–	
γ^1	5	0.15	–	
γ^3	–		1	0.03
δ^1	24	0.74	3	0.09
ε	5	0.15	–	
ζ	7	0.22	2	0.06

Ex Ponto (3194 lines)

	A		B	
α^1	1	0.03	–	
β^1	8	0.25	–	
β^2	2	0.06	–	
γ^1	4	0.13	–	
γ^3	–		1	0.03
δ^1	42	1.31	10	0.31
δ^2	1	0.03	–	
ε	14	0.44	–	
ζ	10	0.31	2	0.06

Fasti (4972 lines)

	A		B	
α^4	1	0.02	–	
α^5	1	0.02	1	0.02
β^1	3	0.06	–	
γ^1	2	0.04	–	
γ^3	–		1	0.02
δ^1	30	0.60	10	0.20
δ^2	1	0.02	–	
ε	10	0.20	–	
ζ	11	0.22	2	0.04

Heroides 1–14 (2190 lines)

	A		B	
α^1	1	0.05	–	
β^1	8	0.37	–	
γ^1	–		1	0.05
δ^1	18	0.82	4	0.19
δ^2	1	0.05	–	
ε	11	0.50	–	
ζ	11	0.50	1	0.05

Heroides 15 (*Epistula Sapphus*) (220 lines)

	A		B	
δ^1	2	0.90	–	
ε	1	0.50	–	

Heroides 16–21 (1564 lines)

	A		B	
α^1	2	0.13	–	
β^1	2	0.13	1	0.06
γ^1	2	0.13	–	
γ^3	–		1	0.06
δ^1	11	0.70	6	0.37
δ^2	1	0.06	–	
ε	11	0.70	–	
ζ	9	0.57	1	0.06

Ibis (644 lines)

	A		B	
β^1	1	0.16	–	
δ^1	10	1.56	1	0.16
ζ	1	0.16	–	

Metamorphoses (11995 lines)

	A		B	
α^1	5	0.04	3	0.03
α^2	2	0.02	1	0.01
α^3	1	0.01		–
α^5	–		2	0.02
α^7	1	0.01		–
β^1	4	0.03		–
γ^1	9	0.07	3	0.03
γ^3	–		1	0.01
δ^1	63	0.53	37	0.31
δ^2	3	0.03		–
ε	22	0.20	8	0.07
ζ	25	0.21	11	0.09

Tristia (3530 lines)

	A		B	
α^2	2	0.06		–
β^1	9	0.25		–
γ^1	1	0.03		–
γ^3	–		1	0.03
δ^1	34	0.96	8	0.23
ε	10	0.28		–
ζ	12	0.34	8	0.23

Nux (182 lines)

	A		B	
β^1	1	0.16		–
δ^1	3	0.48		–
ε	1	0.16		–

Halieutica (134 lines)

	A		B	
α^3	1	0.75		–

From first to last avoidance of alphas, especially α^1s, is consistent and conspicuous.[22] In the early and late works the omega count is notably high, but lower in between (*Ars, Remedia, Fasti, Metamorphoses*). Given the quantity of material, coincidence seems unlikely.

The possibility of scribal inversion in some cases is apparent; e.g. it is hard to feel confident that Ovid wrote *cum pulchrae dominae nostri placuere libelli* in *Am.* 3.8.5 when he could so easily have avoided the α^1, the only one in this work, with *nostri dominae*. Still, in a perhaps unconscious procedure accidental aberrations have their place.

Amores

α^1 A

3.8.5	cum *pulchrae dominae* nostri placuere libelli

β^1 A

2.2.46	ante *suos annos* occidit; illa dea est
2.10.20	simque *mei lecti* non ego solus onus

---

[22] The slightly higher frequency of alphas in the *Metamorphoses* might be attributed to the fact that the elegiac couplet admits them only in the hexameter and the first half of the pentameter. However, the corresponding figures for ζs are a warning not to exaggerate this factor.

| 3.3.14 | perque *meos oculos:* et doluere mei |
| 3.11.48 | perque *tuos oculos,* qui rapuere meos |

β² A

| 1.7.65 | nec *nostris oculis* nec nostris parce capillis |

γ² A

| 1.15.17 | dum fallax *servus, durus* pater, improba lena |

δ¹ A

1.3.8	nomina, *si nostri* sanguinis auctor eques
1.4.60	separor *a domina* nocte iubente mea
1.4.68	si minus, at *certe te* iuvet inde nihil
1.5.19	*quos umeros,* quales vidi tetigique lacertos
1.7.43	denique *si tumidi* ritu torrentis agebar
2.1.16	quod bene *pro caelo* mitteret ille suo
2.2.23	si faciet *tarde, ne te* mora longa fatiget
2.5.29	'quid facis?' *exclamo,* 'quo nunc mea gaudia differs'
2.7.21	*quis Veneris* famulae conubia liber inire
2.13.6	est mihi *pro facto* saepe quod esse potest
2.15.7	felix, *a domina* tractaberis, anule, nostra
2.16.27	quod *si Neptuni* ventosa potentia vincit
2.16.42	separor *a domina* cur ego saepe mea
2.17.27	sunt mihi *pro magno* felicia carmina censu
2.19.18	oscula, *di magni,* qualia quotque dabat
2.19.28	non esset *Danae de* Iove facta parens
3.1.33	altera, *si memini,* limis subrisit ocellis
3.2.21	tu tamen *a dextra,* quicumque es, parce puellae
3.7.53	*a tenera* quisquam sic surgit mane puella
3.8.65	o *si neglecti* quisquam deus ultor amantis
3.10.21	illic, sideream *mundi qui* temperat arcem

δ¹ B

1.8.69	parcius exigito *pretium, dum* retia tendis
1.13.37	*illum dum* refugis, longo quia grandior aevo
1.15.3	non me more *patrum, dum* strenua sustinet aetas
2.2.45	*dum nimium* servat custos Iunonius Ion
2.18.1	carmen ad *iratum dum* tu perducis Achillen
2.18.8	'me *miseram! iam* te' dixit 'amare pudet'

δ² A

| 3.8.8 | turpiter *huc illuc* ingeniosus eo |

ε A

1.4.23	si quid erit *de me* tacita quod mente queraris
1.13.35	Tithono vellem *de te* narrare liceret
1.13.45	ipse deum genitor, *ne te* tam saepe videret
2.2.23	si faciet *tarde, ne te* mora longa fatiget
2.4.31	ut taceam *de me,* qui causa tangor ab omni
3.3.15	dicite, *di, si* vos impune fefellerat illa
3.6.65	*ne me* sperne, precor, tantum, Troiana propago

ζ A

2.2.51	crede *mihi, nulli* sunt carmina grata marito
2.2.57	viderit ipse *licet, credet* tamen ille neganti
2.7.3	sive ego *marmorei respexi* summa theatri
2.9.11	nos tua *sentimus, populus* tibi deditus, arma
3.3.41	quid queror et *toto facio* convicia caelo
3.14.47	prona *tibi vinci* cupientem vincere palma est

ζ B

1.6.15	te *nimium lentum* timeo, tibi blandior uni
1.6.35	hunc ego, si *cupiam, nusquam* dimittere possum
2.2.35	sed tamen *interdum tecum* quoque iurgia nectat
2.11.43	primus ego *aspiciam notam* de litore puppim
3.8.41	nec valido *quisquam terram* scindebat aratro

Ars amatoria, Remedia amoris, De medicamine facie

a^2 A

Ars 2.594	*insidias illas, quas* tulit ipsa, dare

β^1 A

Med. fac. 50	perque *suos annos* hinc bene pendet amor
Rem. 644	deque *tua domina* multa querenda refers

γ^1 A

Ars 2.465	quae modo *pugnarunt, iungunt* sua rostra columbae
Ars 3.435	quae vobis *dicunt, dixerunt* mille puellis
Ars 3.533	carmina qui *facimus, mittamus* carmina tantum
Ars 3.658	saepe *dabit, dederit* quas semel, ille manus
Rem. 49	sed quaecumque *viris, vobis* quoque dicta, puellae

γ^3 B

Ars 2.45	*remigium volucrum* disponit in ordine pinnas

δ¹ A

Ars 1.139	proximus *a domina,* nullo prohibente, sedeto
Ars 1.252	consule *de facie* corporibusque diem
Ars 1.356	quod petis, ex *facili, si* volet illa, feres
Ars 1.510	abstulit, *a nulla* tempora comptus acu
Ars 1.646	sunt genus: in *laqueos, quos* posuere, cadant
Ars 1.694	pensa quid in *dextra, qua* cadet Hector, habes
Ars 2.79	iam Samos *a laeva* (fuerant Naxosque relictae
Ars 2.162	nil opus est *illi qui* dabit arte mea
Ars 2.210	ipse fac in *turba, qua* venit illa, locum
Ars 2.228	tunc quoque *pro servo,* si vocat illa, veni
Ars 2.236	mollibus *his castris* et labor omnis inest
Ars 2.260	ianitor et *thalami qui* iacet ante fores
Ars 2.290	hanc tamen *a domina* fac petat ille tua
Ars 2.359	dum Menelaus abest, *Helene, ne* sola iaceret
Ars 2.421	candidus *Alcathoi qui* mittitur urbe Pelasga
Ars 2.594	*insidias illas quas* tulit ipsa dare
Ars 2.722	ut sol *a liquida* saepe refulget aqua
Ars 3.226	aptius *a summa* conspiciere manu
Ars 3.308	nuda sit, *a laeva* conspicienda manu
Rem. 37	*his lacrimis* contentus eris sine crimine mortis
Rem. 324	*pro vitio* virtus crimina saepe tulit
Rem. 404	desinat; *a prima* proxima segnis erit
Rem. 501	deceptum *risi, qui* se simulabat amare
Rem. 608	laese vir *a domina,* laesa puella viro

δ¹ B

Ars 2.131	ille levi virga (*virgam nam* forte tenebat)
Rem. 473	*quam postquam* reddi Calchas, ope tutus Achillis
Rem. 535	sed bibe plus *etiam quam* quod praecordia poscunt

ε A

Ars 1.371	tum *de te* narret, tum persuadentia verba
Ars 1.484	quaeque roget *ne se* sollicitare velis
Ars 2.393	et, *ne te* capiat latebris sibi femina notis
Ars 2.445	fac timeat *de te,* tepidamque recalface mentem
Rem. 267	omnia fecisti *ne te* ferus urgeat ignis

ζ A

Ars 1.697	forte erat in *thalamo virgo* regalis eodem
Ars 2.194	nec *iubeo collo* retia ferre tuo

Ars 2.261	nec dominam *iubeo pretioso* munere dones
Ars 2.659	si paeta est, *Veneris similis:* si flava, Minervae
Ars 3.467	fert *animus propius* consistere: supprime habenas
Ars 3.605	cum *melius foribus* possis, admitte fenestra
Rem. 213	tu tantum *quamvis firmis* retinebere vinclis

ζ B

Ars 3.715	*iamiam venturam,* quaecumque erat Aura, putabas
Rem. 783	nam sibi quod *numquam tactam* Briseida iurat

Ex Ponto

α¹ A

1.9.4	*invitis oculis* littera lecta tua est

α² A

4.10.55	quique *duas terras,* Asiam Cadmique sororem[23]

β¹ A

1.2.26	cumque *meo fato* quarta fatigat hiems
1.9.7	ante *meos oculos* tamquam praesentis imago
1.9.20	cumque *meis lacrimis* miscuit usque suas
2.5.69	utque *meis numeris* tua dat facundia nervos
2.8.70	utque *meas aquilas,* ut mea signa sequar
2.9.60	atque *suis numeris* forte quievit opus
2.10.44	ante *tuos oculos,* ut modo visus, ero
4.10.20	parte *meae vitae,* si modo dentur, emam

β² A

2.4.7	ante *oculos nostros* posita est tua semper imago
2.8.17	quid *nostris oculis* nisi sola Palatia desunt

γ¹ A

2.8.41	sic pater in *Pylios, Cumaeos* mater in annos
3.3.33	forsitan *exiguas, aliquas* tamen, arcus et ignes
3.8.3	dignus es *argento, fulvo* quoque dignior auro
4.10.76	vir *tanto quanto* debuit ore cani

γ³ B

2.8.26	*terrarum dominum* quem tua cura facit

---

[23] Suspected verse, not counted.

δ¹ A

1.1.45	en ego *pro sistro* Phrygiique foramine buxi
1.1.77	hoc mihi *si superi,* quorum sumus omnia, credent
1.1.80	plus isto, *duri, si* precer, oris ero
1.2.1	Maxime, *qui tanti* mensuram nominis imples
1.2.108	ossa nec *a Scythica* nostra premantur humo
1.2.131	ille ego, *qui duxi* vestros Hymenaeon ad ignes
1.3.2	qui miser est, *ulli si* suus esse potest
1.3.62	*qui forti* casum mente tulere refer
1.4.24	quam laudem *a sera* posteritate ferat
1.4.44	perstiterit *laesi si* gravis ira dei
1.4.56	*dis veris,* memori debita ferre manu
1.5.16	me quoque, *qui feci,* iudice digna lini
1.5.33	*qui, sterili* totiens cum sim deceptus ab aevo
1.6.19	quae, non *mendaci si* quicquam credis amico
1.7.23	nec tamen *irrumpo quo* non licet ire, satisque est
1.8.4	summa satis *nostri si* tibi nota mali
1.10.21	is quoque, *qui gracili* cibus est in corpore, somnus
2.2.9	non ego *concepi, si* Pelion Ossa tulisset
2.3.13	ipse decor, recte *facti si* praemia desint
2.3.44	*a Stygia* quantum mors mea distat aqua
2.3.49	si bene te *novi, si qui* prius esse solebas
2.3.95	pro quibus *optandi si* nobis copia fiat
2.5.12	quod tamen *optari, si* sciat, ipse sinat
2.7.23	crede *mihi, si* sum veri tibi cognitus oris
2.7.33	quae tibi *si memori* coner perscribere versu
2.8.32	moribus *agnosci qui* tuus esse potest
2.8.57	felices *illi, qui* non simulacra, sed ipsos
2.9.26	victima *pro templo* cur cadat icta Iovis
2.10.47	te tamen intueor *quo solo* pectore possum
3.2.23	sint *hi contenti* venia, iactentque, licebit
3.3.57	quid tamen hoc prodest, *vetiti si* lege severa
3.3.74	nec potes *a culpa* dicere abesse tua
3.5.33	namque ego, *qui perii* iampridem, Maxime, vobis
3.5.38	ecquid ab *his ipsis* admoneare mei
3.8.21	hos habet haec *calamos, hos* haec habet ora libellos
4.4.17	consule *Pompeio, quo* non tibi carior alter
4.5.27	cum tamen *a turba* rerum requieverit harum
4.8.90	numina, *pro socero* paene precare tuo
4.9.55	*interea, qua* parte licet, ne cuncta queramur
4.9.63	sic tu *bis fueris* consul, bis consul et ille

4.13.17	nec te *mirari, si* sint vitiosa, decebit
4.13.27	esse parem virtute *patri, qui* frena rogatus

δ¹ B

1.1.62	estque pati *poenam, quam* meruisse, minus
1.2.33	ille ego *sum, lignum* qui non admittar in ullum
1.10.42	Caesaris *offensum dum* mihi numen erit
2.8.67	*quam caream* raptis, o publica numina, vobis
2.9.46	*tam numquam* facta pace cruoris amans
3.2.75	dumque parat *sacrum, dum* velat tempora vittis
3.3.79	haec loca *tum primum* vidi, cum matre rogante
3.5.17	*nam, quamquam* sapor est allata dulcis in unda
4.6.49	*quam quisquam* vestrum qui me doluistis ademptum
4.14.43	tam felix *utinam quam* pectore candidus essem

δ² A

1.7.58	*hic illic* vestro sub lare semper eram

ε A

1.1.29	si dubitas *de me,* laudes admitte deorum
1.3.89	sed vereor *ne me* frustra servare labores
1.8.13	Caspius Aegisos, *de se* si credimus ipsis
2.2.65	exiguam *ne me* praedam sinat esse Getarum
2.3.49	si bene te *novi, si qui* prius esse solebas
2.3.52	utque decet, *ne te* vicerit illa caves
2.5.33	*qui si* forte liber vestras pervenit ad auris
3.2.22	utque habeant *de me* crimina nulla, favet
3.4.88	alter enim *de te,* Rhene, triumphus adest
3.5.43	utque loqui multum *de me* praesente solebas
3.6.6	appellent *ne te* carmina nostra rogas
3.7.31	cur aliquid *de me* speravi lenius umquam
4.9.93	sic ego sum longe, *sic hic,* ubi barbarus hostis
4.9.132	quae *de te* misi caelite facta novo

ζ A

1.5.70	sic *merui, magni* sic voluere dei
2.5.30	nec *potui coepti* pondera ferre mei
2.8.35	parte *leva minima* nostras et contrahe poenas
3.1.107	aemula *Penelopes fieres,* si fraude pudica
3.2.78	sacra *suo facio* barbariora loco
3.5.49	hac ubi *perveni nulli* cernendus in urbem
3.7.3	taedia *consimili fieri* de carmine vobis

4.7.7	ipse vides *certe glacie* concrescere Pontum
4.11.3	tu quoque enim, *memini, caelesti* cuspide facta
4.13.25	nam patris *Augusti docui* mortale fuisse

ζ B

4.3.1	conquerar an *taceam? ponam* sine nomine crimen
4.6.13	iam timeo *nostram cuiquam* mandare salutem

Fasti

α⁴ A

2.287	ipse *deus nudus* nudos iubet ire ministros

α⁵ A

2.357	veste *deus lusus* fallentes lumina vestra

α⁵ B

3.373	ecce levi *scutum versatum* leniter aura

β¹ A

4.112	proque *sua causa* quisque disertus erat
4.848	'sicque *meos muros* transeat hostis' ait
5.58	inque *suo pretio* ruga senilis erat

γ¹ A

1.553	dira viro *facies, vires* pro corpore, corpus
5.607	illa iubam *dextra, laeva* retinebat amictus

γ³ B

5.106	*Pleiadum numerum,* fila dedisse lyrae

δ¹ A

1.115	accipe, *quaesitae quae* causa sit altera formae
1.486	pectora *pro facto* spemque metumque suo
2.343	inde *tori, qui* iunctus erat, velamina tangit
2.462	tum cum *pro caelo* Iuppiter arma dedit
2.559	nec tibi *quae cupidae* matura videbere matri
2.727	dum *nos sollicitos* pigro tenet Ardea bello
3.37	Martia picus avis *gemino pro* stipite pugnant
3.169	cum *sis officiis,* Gradive, virilibus aptus
3.183	quae fuerit, *nostri, si* quaeris, regia nati

3.248	hac sunt, *si memini,* publica facta die
3.506	ei mihi, *pro caelo* qualia dona fero
3.712	Scorpios *a prima* parte videndus erit
3.792	hac, *si commemini,* praeteritaque die
3.795	quid dederit *volucri, si* vis cognoscere, caelum
3.803	viscera *qui tauri* flammis adolenda dedisset
4.203	Iuppiter ortus erat (*pro magno* teste vetustas
4.362	cum tanto *a Phrygia* Gallica distat humus
4.933	dixerat: *a dextra* villis mantele salutis
5.394	'virque' ait '*his armis,* armaque digna viro'
5.418	dicimus, *a media* parte notandus erit
5.646	Albula, *si memini,* tunc mihi nomen erat
5.659	scirpea *pro domino* Tiberi iactatur imago
5.688	nec curent *superi, si* qua locutus ero
6.160	parcite: *pro parvo* victima parva cadit
6.187	quam bene, *di magni,* pugna cecidisset in illa
6.309	venit in *hos annos* aliquid de more vetusto
6.376	Vestaque *pro Latio* multa locuta suo est
6.432	ex *quo iudicio* forma revicta sua est
6.473	iam, Phryx, *a nupta* quereris, Tithone, relinqui
6.552	principiumque *odii, si* sinat illa, canam

$$\delta^1 B$$

3.202	*tum primum* generis intulit arma socer
3.223	qui poterat, clamabat *avum tum* denique visum
3.597	*tum primum* Dido felix est dicta sorori
3.661	haec quoque, *quam referam,* nostras pervenit ad aures
4.404	*tum primum* soles eruta vidit humus
4.525	sic tibi, *quam raptam* quaeris, sit filia sospes
4.615	*tum demum* vultumque Ceres animumque recepit
4.765	neve minus multos *redigam, quam* mane fuerunt
5.540	'*quam nequeam*' dixit 'vincere nulla fera est'
6.291	nec tu aliud *Vestam quam* vivam intellege flammam

$$\delta^2 A$$

| 2.335 | intrat, et *huc illuc* temerarius errat adulter |

$$\varepsilon A$$

1.369	decipiat *ne te* versis tamen ille figuris
1.469	orta prior luna (*de se* si creditur ipsi)
1.593	Africa victorem *de se* vocat, alter Isauras
2.472	*pro quo* nunc dignum sidera munus habent

3.176	hoc solum *ne se* posse Minerva putet
3.233	vel quod erat *de me* feliciter Ilia mater
3.491	praecipue cupiam celari Thesea, *ne te*
3.870	femina, cum *de se* nomina fecit aquae
4.19	si qua tamen pars *te de* fastis tangere debet
6.524	'an numen quod *me, te* quoque vexat?' ait

$$\zeta^1 \text{ A}$$

1.141	ora *vides Hecates* in tres vertentia partes
3.381	plura iubet *fieri simili* caelata figura
3.475	nunc quoque 'nulla *viro' clamabo* 'femina credat'
4.159	templa iubet *fieri Veneri,* quibus ordine factis
4.269	ipsa *peti volui,* nec sit mora, mitte volentem
4.359	'illa deos' *inquit* 'peperit. cessere parenti
5.325	nec *volui fieri* nec sum crudelis in ira
5.411	oscula saepe *dedit, dixit* quoque saepe iacenti
5.657	displicet *heredi mandati* cura sepulcri
6.758	hic, *fieri regni* iura minora sui
6.767	tempora si *veteris quaeris* temeraria damni

$$\zeta \text{ B}$$

4.938	*(quaesieram) 'causam* percipe' flamen ait
6.428	*imperium secum* transferet illa loci

Heroides 1–14

$$\alpha^1 \text{ A}$$

9.124	*invitis oculis*[24] aspicienda venit

$$\beta^1$$

1.102	ille *meos oculos* comprimat, ille tuos
2.91	illa *meis oculis* species abeuntis inhaeret
2.95	cumque *tuis lacrimis* lacrimas coniungere nostras
3.124	cumque *mea patria* laus tua victa iacet
9.121	ante *meos oculos* adducitur advena paelex
12.60	ante *meos oculos* pervigil anguis erat
13.18	sumque *tuos oculos* usque secuta meis
14.68	deque *meis oculis* in tua membra cadunt

---

[24] Recurs in 18.4 and *Ex Pont.* 1.9.4.

γ¹ B

11.3   dextra tenet *calamum, strictum* tenet altera ferrum

δ¹ A

1.66   *quas habitas terras,* aut ubi lentus abes
3.1   quam legis *a rapta* Briseide littera venit
3.82   hic mihi *vae miserae* concutit ossa metus
3.98   at mea *pro nullo* pondere verba cadunt
4.74   *pro rigido* Phaedra iudice fortis erat
5.74   illuc *has lacrimas* in mea saxa tuli
6.108   quaerat et *a patria* Phasidis usque virum
6.125   *legatos quos* paene dedi pro matre ferendos
7.112   prosequitur *fati, qui* fuit ante, tenor
7.123   mille procis *placui, qui* me coiere querentes
7.153   si tibi mens avida est *belli, si* quaerit Iulus
8.15   at tu, cura *mei si* te pia tangit, Oreste
8.26   aspera *pro caro* bella tulisse toro
8.63   *has solas* habeo semper semperque profundo
8.109   *pro somno* lacrimis oculi funguntur obortis
12.133   ausus es—o, *iusto* desunt sua verba dolori
13.128   *a patria* pelago vela vetante datis
14.65   quid mihi cum *ferro? quo* bellica tela puellae

δ¹ B

3.16   infelix *iterum sum* mihi visa capi
3.123   an *tantum, dum* me caperes, fera bella probabas
10.74   te fore, *dum nostrum* vivet uterque, meam
12.202   dos mea, *quam, dicam* si tibi 'redde', neges

δ² A

13.34   creditur, *huc illuc,* qua furor egit, eo

ε A

1.60   ille mihi *de te* multa rogatus abit
3.5   si mihi pauca queri *de te* dominoque viroque
3.77   exagitet *ne me* tantum tua, deprecor, uxor
3.147   me petat ille tuus, *qui, si* dea passa fuisset
5.4   laesa queror *de te,* si sinis, ipsa meo
6.9   cur mihi fama prior *de te* quam littera venit
9.48   et mater *de te* quaelibet esse potest
12.22   hac fruar: haec *de te* gaudia sola feram

12.132	*pro quo* sum totiens esse coacta nocens
12.193	redde torum, *pro quo* tot res insana reliqui
14.98	et, *te ne* feriant, quae geris arma, times

ζ A

1.66	*quas habitas terras,* aut ubi lentus abes
1.71	quid timeam *ignoro–timeo* tamen omnia demens
2.20	ipsa *mihi dixi:* 'si valet ille, venit'
3.13	*differri potui;* poenae mora grata fuisset
4.152	certa *fui, certi si* quid haberet amor
8.3	quod *potui, renui,* ne non invita tenerer
8.119	per patris ossa *tui, patrui* mihi, quae tibi debent
12.55	tristis *abis; oculis* absentem prosequor udis
12.169	non mihi grata *dies; noctes* vigilantur amarae
13.135	sed quid *ago? revoco?* revocaminis omen abesto
14.124	quaeque *tibi tribui* munera, dignus habes

ζ B

10.46	*postquam desieram* vela videre tua

15 *(Epistula Sapphus)*

δ¹ A

15.87	hunc ne *pro Cephalo* raperes, Aurora, timebam
15.115	non aliter quam *si nati* pia mater adempti

ε A

15.103	nil *de te* mecum est nisi tantum iniuria; nec tu

16–21

α¹ A

18.4	*invitis oculis* haec mea verba leges
21.21	mox ubi, *secreti longi* causa optima, somnus

β¹ A

21.27	inde *meos digitos* iterum repetita fatigat
21.82	inque *meis oculis* candida Delos erat

β¹ B

20.212	esse *tuam vinctam* numine teste fidem

γ¹ A

17.127	sed nihil *infirmo; faveo* quoque laudibus istis
21.143	non ego *iuravi, legi* iurantia verba

γ³ B

16.94	*multarum votum* sola tenere potes

δ¹ A

16.142	nullaque *de facie* nescia terra tua est
16.172	deprecor, *o tanto* digna labore peti
16.287	a nimium simplex *Helene, ne* rustica dicam
16.293	vix *fieri, si* sunt vires in semine morum
16.341	nec tu rapta *time ne* nos fera bella sequantur
17.48	error, *qui facti* crimen obumbret, erit
17.173	*de facie* metuit, vitae confidit, et illum
18.27	his ego *si vidi* mulcentem pectora somnum
18.102	oscula, *di magni*, trans mare digna peti
20.116	mitis adhuc *fieri, si* patiare, potest
21.169	at mihi *vae miserae* torrentur febribus artus

δ¹ B

17.201	dum loqueris *mecum, dum* nox sperata paratur
17.202	qui ferat in *patriam, iam* tibi ventus erit
19.105	a potius *peream quam* crimine vulnerer isto
20.207	et te, *dum nimium* miror, nota certa furoris
21.107	mittitur ante pedes *malum cum* carmine tali
21.174	daque *salutiferam iam* mihi fratris opem

δ² A

20.130	anxius *huc illuc* dissimulanter eo

ε A

16.141	magna quidem *de te* rumor praeconia fecit
16.325	si pudet et metuis *ne me* videare secuta
16.375	tu quoque, si *de te* totus contenderit orbis
17.18	et laudem *de me* nullus adulter habet
17.197	nec tamen ipse negas, et nobis omnia *de te*
17.209	quid *de me* poterit Sparte, quid Achaia tota
17.211	quid Priamus *de me*, Priami quid sentiet uxor
19.19	aut ego cum cara *de te* nutrice susurro
20.94	quod *de me* solo nempe queraris habes

| 20.197 | non agitur *de me;* cura maiore laboro. |
| 21.193 | iam quoque nescio quid *de me* sensisse videtur |

ζ A

16.318	in *viduo iaceo* solus et ipse toro
17.143	nunc quoque, quod *tacito mando* mea verba libello
17.162	nil *illi potui* dicere praeter 'erit'
18.143	*invideo Phrixo,* quem per freta tristia tutum
18.175	an malim *dubito, toto* procul orbe remotus
19.157	in tua castra *redi, socii* desertor amoris
21.9	utque cupis *credi, memori* te vindicat ira
21.102	quidquid *ibi vidi* dicere—Delos habet
21.216	in *pomo refero* mente fuisse tuo

ζ B

| 16.95 | nec *tantum regum* natae petiere ducumque |

Ibis

β¹ A

| 321 | inque *tuo thalamo* ritu iugulere Pheraei |

δ¹ A

29	et tibi, *calcasti qui* me, violente, iacentem
52	teque *brevi, qui* sis, dissimulare sinam
128	Phoebus, et *a laeva* maesta volavit avis
281	vel quae *qui redimi* Romano turpe putavit
287	aut velut *Eurylochi, qui* sceptrum cepit ab illo
359	filia si *fuerit, sit* quod Pelopea Thyesti
381	ut *qui Threicii* quondam praesepia regis
392	inque caput *domini qui* dabat arma procis
477	praedaque *sis illis,* quibus est Latonia Delos
523	utque parum *stabili qui* carmine laesit Athenas

δ¹ B

| 71 | ipsaque tu, Tellus, *ipsum cum* fluctibus aequor |

ζ A

| 117 | sisque *miser semper,* nec sis miserabilis ulli |

Metamorphoses

α^1 A

3.66	quod medio *lentae spinae* curvamine fixum
7.155	somnus in *ignotos oculos* sibi venit, et auro
7.559	sed *dura terra* ponunt praecordia, nec fit
12.570	*infirmis pennis,* et qua levis haeserat alae
15.364	in scrobe *delectos mactatos* obrue tauros

α^1 B

6.104	Europam: *verum taurum,* freta vera putares
11.138	perque *iugum Lydum* labentibus obvius undis
14.745	edidit et *matrum miserarum* facta peregit

α^2 A

5.26	ex *illis scopulis,* ubi erant affixa, petisses
13.642	bisque *duas natas,* quantum reminiscor, habebas

α^2 B

12.187	spectatorem *operum multorum* reddere, vixi

α^3 A

8.30	*imposito calamo* patulos sinuaverat arcus

α^5 B

7.429	colla torosa *boum vinctorum* tempora vittis
15.381	fingit et in *formam, quantam* capit ipsa, reducit

α^7 A

1.266	barba gravis *nimbis, canis* fluit unda capillis

β^1 A

2.297	vixque *suis umeris* candentem sustulit axem
7.168	deme *meis annis* et demptos adde parenti
11.304	ille *suis Delphis,* hic vertice Cylleneo
13.130	tuque *tuis armis,* nos te poteremur, Achille

γ^1 A

1.292	omnia pontus *erant, deerant* quoque litora ponto
1.552	ora cacumen *habet: remanet* nitor unus in illa
2.322	etsi non *cecidit, potuit* cecidisse videri
4.756	alipedi *vitulus, taurus* tibi, summe dearum
11.152	Sardibus *hinc, illinc* parvis finitur Hypaepis

11.197	dextera *Sigei, Rhoetaei* laeva profundi
13.791	splendidior *vitro, tenero* lascivior haedo
15.502	quod *voluit, finxit* voluisse et, crimine verso
15.568	quae *vidit, tetigit,* nec iam sua lumina damnans

γ¹ B

2.739	Aglauros *laevum, medium* possederat Herse
9.147	conquerar an *sileam? repetam* Calydona morerne
14.128	templa tibi *statuam, tribuam* tibi turis honores

γ³ B

10.16	*umbrarum dominum* pulsisque ad carmina nervis

δ¹ A

1.297	figitur in *viridi, si* fors tulit, ancora prato
1.581	moxque amnes *alii, qui,* qua tulit impetus illos
1.589	flumine et 'o *virgo,* Iove digna tuoque beatum
2.25	*a dextra* laevaque Dies et Mensis et Annus
2.323	quem procul *a patria* diverso maximus orbe
2.462	*qua posita* nudo patuit cum corpore crimen
2.562	acta deae *refero. pro quo* mihi gratia talis
2.699	'rustice, *vidisti si* quas hoc limite' dixit
2.735	ut teres in *dextra, qua* somnos ducit et arcet
2.870	cum deus *a terra* siccoque a litore sensim
3.161	fons sonat *a dextra* tenui perlucidus unda
3.525	meque sub *his tenebris* nimium vidisse quereris
3.632	'quid *facitis? quis* clamor?' ait. 'qua, dicite, nautae
3.655	si puerum iuvenes, *si multi* fallitis unum
4.111	in loca plena metus *qui iussi* nocte venires
4.649	huic quoque 'vade procul *ne longe* gloria rerum
4.788	quae freta, *quas terras* sub se vidisset ab alto
4.797	pars fuit: *inveni, qui* se vidisse referret
5.108	*invicti, vinci si* possent caestibus enses
5.252	deserit, *a dextra* Cythno Gyaroque relictis
5.378	at tu *pro socio,* si qua est ea gratia, regno
5.436	denique *pro vivo* vitiatas sanguine venas
5.489	atque ait 'o *toto* quaesitae virginis orbe
5.628	aut *lepori, qui* vepre latens hostilia cernit
6.224	e quibus Ismenus, *qui matri* sarcina quondam
6.656	atque ubi *sit quaerit:* quaerenti iterumque vocanti
7.357	Aeoliam Pitanen *a laeva* parte relinquit
7.482	arma iuves *oro pro nato* sumpta piaeque

7.499	*a dextra* laevaque duos aetate minores
7.590	*pro nato* genitor dum verba precantia dicit
7.615	'Iuppiter, o' *dixi* 'si te non falsa loquuntur
8.58	iusta gerit certe *pro nato* bella perempto
8.129	me perimat. cur, *qui vicisti* crimine nostro
8.131	officium tibi sit. *te vere* coniuge digna est
8.190	*a minima* coeptas, longam breviore sequenti
9.82	induit ille toris *a laeva* parte lacertos
9.441	et Minoa, *queri. qui* dum fuit integer aevi
9.667	implesset *monstri, si* non miracula nuper
11.168	sustinet *a laeva*, tenuit manus altera plectrum
11.386	disicit *hos ipsos* colloque infusa mariti
11.439	quod tua *si flecti* precibus sententia nullis
11.450	hoc quoque lenimen, *quo solo* flexit amantem
11.548	verum ubi *sit nescit:* tanta vertigine pontus
11.583	at dea non ultra *pro functo* morte rogari
11.660	inveniesque *tuo pro* coniuge coniugis umbram
12.248	elatumque alte, *veluti qui* candida tauri
12.336	et missum *a dextra* laevam penetravit ad aurem
12.446	tum poteram *magni, si* non superare, morari
13.23	moenia *qui forti* Troiana sub Hercule cepit
13.113	quod tibi *si populi* donaverit error Achivi
13.115	et fuga, *qua sola* cunctos, timidissime, vincis
13.138	quae nunc *pro domino, pro* vobis saepe locuta est
13.214	mente ferant placida, *doceo, quo* simus alendi
13.284	*his umeris, his*, inquam, umeris ego corpus Achillis
13.288	scilicet *idcirco pro nato* caerula mater
13.326	quam, cessante *meo pro* vestris pectore rebus
13.596	*pro patruo* tulit arma suo primisque sub annis
13.695	*pro populo* cecidisse suo pulchrisque per urbem
13.864	sentiet esse mihi *tanto pro* corpore vires
14.56	inquinat; *his fusis* latices radice nocenti
14.356	si modo me *novi, si* non evanuit omnis
14.444	ereptam *Argolico quo* debuit igne cremavit
15.464	impius humano, *vituli qui* guttura ferro

δ¹ B

1.119	*tum primum*[25] siccis aer fervoribus ustus
1.185	*nam quamquam* ferus hostis erat, tamen illud ab uno

[25] Repeated 121 and 123.

2.57	plus *etiam, quam* quod superis contingere possit
2.171	*tum primum* radiis gelidi caluere Triones
2.257	fors eadem Ismarios *Hebrum cum* Strymone siccat
2.636	filia centauri, *quam quondam* nympha Chariclo
3.95	*dum spatium* victor victi considerat hostis
3.230	Actaeon ego *sum: dominum* cognoscite vestrum
3.342	caerula Liriope, *quam quondam* flumine curvo
4.83	ad solitum coiere *locum. tum* murmure parvo
4.283	et Crocon in parvos *versum cum* Smilace flores
4.542	Leucothoeque *deum cum* matre Palaemona dixit
5.556	*quam postquam* toto frustra quaesistis in orbe
6.3	*tum secum:* 'laudare parum est, laudemur et ipsae
6.225	prima suae fuerat, *dum certum* flectit in orbem
6.402	vulgus et *exstinctum cum* stirpe Amphiona luget
7.297	neve doli cessent, *odium cum* coniuge falsum
7.627	*dum numerum* miror, 'totidem, pater optime', dixi
7.675	'sum nemorum studiosus' ait 'caedisque ferinae
8.24	plus *etiam quam* nosse sat est: hac iudice Minos
8.477	impietate pia est. *nam postquam* pestifer ignis
8.757	dixit, et *obliquum dum telum* librat in ictus
9.413	*tum demum* magno petet hos Acheloia supplex
9.604	plura loqui *poteram, quam* quae cepere tabellae
9.669	proxima Cnosiaco *nam quondam* Phaestia regno
10.318	elige, Myrrha, *virum, dum* ne sit in omnibus unus
11.66	Eurydicenque *suam, iam* tuto respicit Orpheus
12.5	postmodo qui rapta *longum cum* coniuge bellum
13.303	haut timeo, si *iam nequeam* defendere, crimen
13.391	dixit et in pectus *tum demum* vulnera passum
13.960	hanc ego *tum primum* viridi ferrugine barbam
13.961	caesariemque *meam, quam* longa per aequora verro
14.450	non sine Marte tamen. *bellum cum* gente feroci
14.576	congerie e media *tum primum* cognita praepes
14.725	*quam vitam* geminaque simul mihi luce carendum
15.534	reddita vita foret; *quam postquam* fortibus herbis
15.589	exul *agam, quam* me videant Capitolia regem

$$\delta^2 \, A$$

6.365	*huc illuc* limum saltu movere maligno
7.581	*hic illic,* ubi mors deprenderat exhalantes
12.329	et quatit *huc illuc* labefactaque robora iactat

ε A

2.75	obvius ire polis, *ne te* citus auferat axis
2.562	acta deae *refero. pro quo* mihi gratia talis
2.571	divitibusque procis (*ne me* contemne) petebar
3.117	exclamat '*ne te* civilibus insere bellis'
4.44	cogitat et dubia est, *de te,* Babylonia, narret
4.584	dumque aliud superest *de me,* me tange manumque
5.201	miles erat Persei: *pro quo* dum pugnat, Aconteus
7.32	hoc ego si patiar, tum *me de* tigride natam
7.146	sed *te, ne* faceres, tenuit reverentia famae
7.854	per si quid merui *de te* bene perque manentem
8.60	et puto, vincemur, *qui, si* manet exitus urbem
9.578	'effuge' ait. '*qui, si* nostrum tua fata pudorem
9.728	cura tenet Veneris? *si di* mihi parcere vellent
10.239	esse negare deam; *pro quo* sua numinis ira
10.274	constitit et timide '*si di* dare cuncta potestis
10.298	editus hac ille est, *qui, si* sine prole fuisset
11.695	et *ne me* fugeres, ventos sequerere, rogabam
13.49	quae meruit, quae, *si di* sunt, non vana precaris
14.588	Aeneaeque meo, qui *te de* sanguine nostro
14.678	Vertumnumque tori socium tibi selige. *pro quo*
15.742	huc *se de* Latia pinu Phoebeius anguis
15.816	hic sua complevit, *pro quo,* Cytherea, laboras

ε B

8.145	quam pater ut vidit (*nam iam* pendebat in aura)
9.388	plura loqui nequeo. *nam iam* per candida mollis
9.403	non est passa Themis: '*nam iam* discordia Thebae
12.205	sicut erat: *nam iam* voto deus aequoris alti
13.450	rapta sinu matris, *quam iam* prope sola fovebat
14.144	quae patienda diu est. *nam iam* mihi saecula septem
14.399	inveniunt Circen (*nam iam* tenuaverat auras)
14.750	venit Anaxaretes, *quam iam* deus ultor agebat

ζ A

1.465	cuncta, *deo, tanto* minor est tua gloria nostra
2.178	ut *vero summo* dispexit ab aethere terras
2.382	ipse sui *decoris, qualis,* cum deficit orbem
4.765	postquam epulis *functi generosi* munera Bacchi
5.108	*invicti, vinci si* possent caestibus enses
6.207	et nisi *Iunoni nulli* cessura deorum
7.837	'aura *veni' dixi* 'nostroque medere dolori'

9.8   nomine si qua *suo fando* pervenit ad aures
9.302   nitor, et *ingrato facio* convicia demens
9.476   ille quidem est *oculis quamvis* formosus iniquis
9.490   omnia di *facerent essent* communia nobis
10.25   posse *pati volui* nec me temptasse negabo
10.685   *exemplo caveo* meque ipsa exhortor in ambos
11.222   concipe. mater *eris iuvenis,* qui fortibus annis
11.722   fit *propius corpus:* quod quo magis illa tuetur
13.180   arma *peto: vivo* dederam, post fata reposco
13.206   utiliter *feci spatiosi* tempore belli
13.516   inferias *hosti peperi.* quo ferrea resto
14.183   vidi iterum *veluti tormenti* viribus acta
14.273   nec mora, *misceri tosti* iubet hordea grani
14.550   in capitum *facies puppes* mutantur aduncae
14.760   servat adhuc *Salamis, Veneris* quoque nomine templum
15.64   visibus *humanis, oculis* ea pectoris hausit
15.160   ipse ego (nam *memini) Troiani* tempore belli
15.759   humano *generi, superi,* favistis abunde

ζB

1.710   'hoc mihi *concilium tecum*' dixisse 'manebit'
2.197   porrigit in *spatium signorum* membra duorum
3.485   ducere *purpureum nondum* matura colorem
7.299   confugit atque *illam, quoniam* gravis ipse senecta est
8.169   quo *postquam geminam* tauri iuvenisque figuram
9.328   fata meae *referam? quamquam* lacrimaeque dolorque
11.426   non *etiam metuam,* curaeque timore carebunt
12.261   ignibus et *medium Lapitharum* iecit in agmen
12.523   Tartara *detrusum silvarum* mole ferebant
13.462   mors tantum *vellem matrem* mea fallere posset
13.855   hunc tibi do *socerum; tantum* miserere precesque

Tristia

α²A

2.61   quid referam *libros, illos* quoque, crimina nostra
4.8.19   ne cadat et *multas palmas* inhonestet adeptus

β¹A

2.222   inque *tuis umeris* tam leve fertur onus
2.342   inque *meas poenas* ingeniosus eram
3.3.82   deque *tuis lacrimis* umida serta dato

3.14.50	inque *meis scriptis* Pontica verba legas
4.2.61	illa *meos oculos* mediam deducit in urbem
4.4.85	aque *mea terra*[26] prope sunt funebria saxa
5.2.22	parsque *meae poenae* totius instar erit
5.10.10	cumque *meis curis* omnia longa facit
5.12.36	digna *sui domini* tempore, digna loco

$$\gamma^1\,A$$

4.6.37	nos quoque quae *ferimus, tulimus* patientius ante

$$\gamma^3\,B$$

5.11.19	utque aliis, *quorum numerum* comprendere non est

$$\delta^1\,A$$

1.1.88	ut satis *a media* sit tibi plebe legi
1.1.93	si poteris vacuo *tradi, si* cuncta videbis
1.1.128	ultimus, *a terra* terra remota mea
1.2.84	quodque sit *a patria* tam fuga tarda queror
1.2.98	*a culpa* facinus scitis abesse mea
1.2.101	quod licet et minimis, *domui si* favimus illi
1.2.103	hoc duce *si dixi* felicia saecula, proque
1.3.46	*pro deplorato* non valitura viro
1.3.54	horam, *propositae quae* foret apta viae
1.5.66	*a patria* fugi victus et exul ego
2.2	ingenio *perii qui* miser ipse meo
2.46	*vidi qui* tulerant in caput arma tuum
2.56	*qua sola* potui, mente fuisse tuam
2.97	me miserum! *potui, si* non extrema nocerent
2.230	bellaque *pro magno* Caesare Caesar obit
2.303	et procul *a scripta* solis meretricibus Arte
2.456	impresso *fieri qui* solet ore, docet
2.572	gratia *candori si* qua relata meo est
3.10.60	et *quas divitias* incola pauper habet
3.11.45	aspicis *a dextra* latus hoc adapertile tauri
3.14.10	*qui domini* poenam non meruere sui
3.14.22	certius *a summa* nomen habere manu
4.1.32	*illo, quo* nocuit, grata sopore fuit
4.1.68	illum *qui populi* semper in ore fuit
4.2.46	vincula fert *illa qua* tulit arma manu
4.8.43	hoc mihi *si Delphi* Dodonaque diceret ipsa

---

[26] Owen's conjecture for *atque meam terram*.

5.1.41	lenior *invicti si* sit mihi Caesaris ira
5.2.8	affectusque *animi, qui* fuit ante, manet
5.3.11	nunc procul *a patria* Geticis circumsonor armis
5.5.60	implerent *venti si* mea vela sui
5.10.33	hos quoque, *qui geniti* Graia creduntur ab urbe
5.10.34	*pro patrio* cultu Persica braca tegit
5.11.16	*qui merui* vitio perdere cuncta meo
5.12.30	*illi, qui* fueram, posse redire parem

$$\delta^1\,\mathrm{B}$$

1.2.41	o bene, quod non *sum mecum* conscendere passus
3.3.21	si *iam deficiam,* suppressaque lingua palato
3.4.15	*dum tecum* vixi, dum me levis aura ferebat
3.12.5	*iam violam* puerique legunt hilaresque puellae
4.4.30	plus *etiam quam* me iudice dignus eram
4.5.9	temporis *oblitum dum* me rapit impetus huius
4.8.49	*nam quamquam* vitio pars est contracta malorum
5.2.42	ancora *iam nostram* non tenet ulla ratem

$$\varepsilon\,\mathrm{A}$$

1.1.63	clam tamen intrato, *ne te* mea carmina laedant
2.217	*de te* pendentem sic dum circumspicis orbem
2.448	sic etiam *de se* quod neget illa viro
3.7.21	sed vereor *ne te* mea nunc fortuna retardet
3.14.13	Palladis exemplo *de me* sine matre creata
4.4.45	idque deus sentit; *pro quo* nec lumen ademptum
4.7.26	excusem *ne te* semper, amice, mihi
4.10.84	sum miser, et *de me* quod doluere nihil
5.10.39	meque palam *de me* tuto mala saepe loquuntur
5.14.17	quod numquam vox est *de te* mea muta tuique

$$\zeta\,\mathrm{A}$$

1.2.67	est *illi nostri* non invidiosa cruoris
2.30	non *adeo nostro* fugit ab ore pudor
2.463	non fuit hoc *illi fraudi,* legiturque Tibullus
2.563	non ego *mordaci destrinxi* carmine quemquam
3.1.69	altera templa *peto, vicino* iuncta theatro
3.4.78	*contacti simili* sorte rogetis idem
3.6.5	isque erat usque *adeo populo* testatus, ut esset
3.8.9	desertaeque *domus vultus,* memoresque sodales
3.13.21	funeris ara *mihi, ferali* cincta cupressu
4.3.56	et *dici, memini,* iuvit et esse meam

5.1.9	ut *cecidi, subiti* perago praeconia casus
5.10.50	non *merui tali* forsitan esse loco

ζ B

1.2.43	at nunc, ut *peream, quoniam* caret illa periclo
1.5.51	pars *etiam quaedam* mecum moriatur oportet
2.387	tingueret ut *ferrum natorum* sanguine mater
2.575	non ut in *Ausoniam redeam,* nisi forsitan olim
4.1.27	non *equidem vellem,* quoniam nocitura fuerunt
4.1.81	sic si quem *nondum portarum* saepe receptum
4.4.47	forsitan hanc *ipsam, vivam* modo, finiet olim
5.6.35	elige *nostrorum minimum* minimumque laborum

Nux

β¹ A

138	sitque *tuis muris,* Romule iuris idem

δ¹ A

90	non *a vicina* pulverulenta via est
177	*si merui* videorque nocens, imponite flammae
179	*si merui* videorque nocens, excidite ferro

ε A

130	non spolium *de me* quod petat hostis habet

Halieutica

α³ A

36	*elato calamo* cum demum emersus in auras

PANEGYRICUS MESSALLAE (211 lines)

	A		B	
α^4	1	0.47	–	
ε	1	0.47	–	
ζ	–		1	0.47

α^4 A

16   hic quoque sit *gratus parvus* labor, ut tibi possim

ε A

211   inceptis *de te* subtexam carmina chartis

ζ B

1   te, Messalla, *canam, quamquam* tua cognita virtus

PERSIUS (650 lines)

	A		B	
α^1	1	0.15	–	
α^2		–	1	0.15
δ^1	7	1.08	–	
ε	1	0.15	–	
ζ	1	0.15	–	

α^1 A

1.22   tun, vetule, *auriculis alienis* colligis escas

α^2 B

1.29   ten *cirratorum centum* dictata fuisse

δ^1 A

1.30   *pro nihilo* pendes? ecce inter pocula quaerunt
1.103   haec fierent, *si testiculi* vena ulla paterni
2.22   dic agedum *Staio*, 'pro Iuppiter, o bone' clamet
3.72   iussit et *humana qua* parte locutus es in re
4.27   hunc ais, hunc *dis iratis* genioque sinistro
5.69   egerit *hos annos* et semper paulum erit ultra
5.130   nascuntur *domini, qui* tu impunitior exis

$$\varepsilon\,A$$

1.55   et 'verum' inquis 'amo, verum mihi dicite *de me*

$$\zeta\,A$$

1.10   aspexi ac *nucibus facimus* quaecumque relictis

PETRONIUS, *De bello civili* (295 lines)

	A		B	
α^1	1	0.34	–	
δ^1	–		2	0.68
ε	1	0.34	–	
ζ	1	0.34	–	

$$\alpha^1\,A$$

146   est locus *Herculeis aris* sacer. hunc nive dura

$$\delta^1\,B$$

160   pulsus ab urbe mea, *dum Rhenum* sanguine tingo
269   *Magnum cum* Phoebo soror et Cyllenia proles

$$\varepsilon\,A$$

231   id *pro quo* metuit, tantum trahit. omnia secum

$$\zeta\,A$$

162   Alpibus *excludo, vincendo* certior exul

PROPERTIUS

	A		B	
		I (706 lines)		
β^1	1	0.14	–	
δ^1	2	0.30	1	0.14
ε	4	0.57	–	
ζ	2	0.30	–	

	A		B	

II (1362 lines)

	A		B	
α^2	–		1	0.07
α^6	1	0.07	–	
β^1	2	0.15	–	
γ^1	1	0.07	–	
δ^1	9	0.66	1	0.07
ε	10	0.73	1	0.07
ζ	4	0.29	–	

I have excluded 3.7.49 *sed Thyio thalamo aut Oricia terebintho* from the register
(see p. 2 n. 7).

III (990 lines)

	A		B	
α^1	1	0.10	–	
α^2	1	0.10	–	
α^3	1	0.10	–	
β^1	2	0.20	–	
γ^1	1	0.10	–	
δ^1	2	0.20	1	0.10
ε	4	0.40	–	
ζ	3	0.30	–	

IV (952 lines)

	A		B	
α^1	1	0.11	–	
β^1	3	0.31	–	
γ^1	1	0.11	–	
δ^1	2	0.21	3	0.31
ε	4	0.41	–	
ζ	1	0.11	2	0.21

I

β^1 A

13.32   illa *suis verbis* cogat amare Iovem

δ^1 A

10.24   neu tibi *pro vano* verba benigna cadant
20.1   hoc *pro continuo* te, Galle, monemus amore

δ¹B

| 20.48 | *tum sonitum* rapto corpore fecit Hylas |

ε A

8.17	sed quocumque modo *de me,* periura, mereris
8.21	nam me non ullae poterunt corrumpere, *de te*
18.27	*pro quo* continui montes et frigida rupes
19.21	quam vereor *ne te* contempto, Cynthia, busto

ζ A

| 8.40 | sed *potui blandi* carminis obsequio |
| 14.16 | nulla *mihi tristi* praemia sint Venere |

II

α²B

| 10.10 | nunc *aliam citharam* me mea Musa docet |

α⁶A

| 31.5 | hic equidem *Phoebus visus* mihi pulchrior ipso |

β¹A

| 5.18 | parce *tuis animis,* vita, nocere tibi |
| 25.4 | Calve, *tua venia,* pace, Catulle, tua |

γ¹A

| 1.37 | Theseus *infernis, superis* testatur Achilles |

δ¹A

1.49	*si memini,* solet illa levis culpare puellas
11.2	laudet, *qui sterili* semina ponit humo
15.41	qualem *si cuncti* cuperent decurrere vitam
16.42	*illa, qua* vicit, condidit arma manu
23.6	integit? et *campo quo* movet illa pedes
24.3	cui non *his verbis* aspergat tempora sudor
24.48	et se plus *uni si* qua parare potest
25.32	nescio *quo pacto* verba nocere solent
34.14	*a domina* tantum te modo tolle mea

δ¹B

| 25.26 | septima *quam metam* triverit axe rota |

ε A

3.4   et turpis *de te* iam liber alter erit
8.40   mirum, si *de me* iure triumphat Amor
9.22   forsitan et *de me* verba fuere mala
20.13   *de te* quodcumque, ad surdas mihi dicitur aures
20.22   cum de me et *de te* compita nulla tacent
21.1   a quantum *de me* Panthi tibi pagina finxit
21.9   dispeream, si quicquam aliud quam gloria *de te*
27.11   solus amans novit quando periturus et *a qua*
32.21   sed *de me* minus est. famae iactura pudicae
32.23   nuper enim *de te* nostras maledixit ad aures

ε B

15.22   viderit haec, si *quam iam* peperisse pudet

ζ A

12.2   nonne *putas miras* hunc habuisse manus
18.37   credam ego *narranti, noli* committere, famae
30.19   num tu, dure, *paras Phrygias* nunc ire per undas
30.25   mi nemo *obiciat. libeat* tibi, Cynthia, mecum

III

α¹ A

6.35   quae tibi si *veris animis* est questa puella

α² A

6.37   et mea cum *multis lacrimis* mandata reporta

α³ A

14.25   *nullo praemisso* de rebus tute loquaris

β¹ A

8.24   sive *meas lacrimas* sive videre tuas
19.24   Nise, *tuas portas* fraude reclusit Amor

γ¹ A

20.23   at quibus *imposuit, solvit* mox vincla libido

δ¹ A

| 7.12 | nunc tibi *pro tumulo* Carpathium omne mare est |
| 22.2 | Tulle, *Propontiaca qua* fluit Isthmos aqua |

δ¹ B

| 16.9 | peccaram semel, et *totum sum* pulsus in annum |

ε A

7.64	hoc *de me* sat erit, si modo matris erit
12.13	neve aliquid *de te* flendum referatur in urna
25.2	et *de me* poterat quilibet esse loquax

ζ A

7.36	*consenuit, fallit* portus et ipsa fidem
12.9	illa quidem *interea fama* tabescet inani
25.3	quinque *tibi potui* servire fideliter annos

IV

α¹ A

| 11.43 | non fuit *exuviis tantis* Cornelia damnum |

β¹ A

1.80	inque *meis libris* nil prius esse fide
10.47	sed quia victa *suis umeris* haec arma ferebant
11.6	nempe *tuas lacrimas* litora surda bibent

γ¹ A

| 5.9 | illa *velit, poterit* magnes non ducere ferrum |

δ¹ A

| 5.25 | seu *quae palmiferae* mittunt venalia Thebae |
| 7.42 | garrula *de facie* si qua locuta mea est |

δ¹ B

6.69	bella satis cecini: *citharam iam* poscit Apollo
11.93	discite *venturam iam* nunc lenire senectam
11.100	*dum pretium* vitae grata rependit humus

ε A

2.20	*de se* narranti tu modo crede deo
4.65	experiar somnum, *de te* mihi somnia quaeram
11.49	quamlibet austeras *de me* ferat urna tabellas
11.81	sat tibi sint noctes, quas *de me,* Paulle, fatiges

ζ A

6.65	di *melius! quantus* mulier foret una triumphus

ζ B

2.39	pastor me ad *baculum possum* curvare vel idem
5.29	et simulare *virum pretium* facit: utere causis

SILIUS (12182 lines)

	A		B	
α^1	19	0.16	1	0.01
α^2	1	0.01	1	0.01
α^3	6	0.05	–	
α^4	1	0.01	1	0.01
α^5	7	0.06	2	0.02
α^6	4	0.03	–	
β^1	1	0.01	–	
γ^1	1	0.01	1	0.01
γ^3	–		4	0.03
δ^1	51	0.42	36	0.30
δ^2	2	0.02	–	
ε	1	0.01	–	
ζ	9	0.07	9	0.07

The alpha count, very high relatively to the other silver-age epicists, may reflect a lack of sensibility, though the omega count is unremarkable.

α^1 A

2.154	sed fisus *latis umeris* et mole iuventae
2.447	quam circa *immensi populi* condensaque cingunt
3.656	*immensis spatiis,* nisi quem cava nubila torquens
5.501	incumbunt *sociae dextrae,* magnoque fragore
6.337	*saxosis latebris* intento ad proelia circum
6.434	*affixi clipei* currusque et spicula nota

6.585	nec *patrias iras* animosque in proelia ferres
8.42	*antiquae patriae,* mandataque magna sororis
8.287	*turbati vulgi,* signataque mente cicatrix
8.633	*immensis tenebris,* et terram et litora Sipus
10.279	pectora *Romuleo regno.* calcaribus aufer
10.316	*transfixi clipei* galeaeque et inutile ferrum
11.517	*optatae pugnae* castris cita signa movemus
12.220	mallet et *Aonio plectro* mulcere labores
12.673	*Aetolos campos?* ubi cum Tyrrhena natarent
16.213	quanta per *Ausonios populos* torrentibus armis
16.330	*antiqui stabuli;* sunt quos spes grata fatiget
16.539	concurrere *animis quantis* confligere par est
16.646	magnanimis *gemino leto,* cum tota subisset

$$\alpha^1\,B$$

| 5.669 | *caesorum iuvenum* Poenus: quae vulnera cernis |

$$\alpha^2\,A$$

| 11.43 | deprensa et *nulla macula* non illita vita |

$$\alpha^2\,B$$

| 14.285 | *unum naufragium,* mersasque impune profundo |

$$\alpha^3\,A$$

5.500	*deposito clipeo* mutatus tela, Sychaeus
9.121	*audita patria* natisque et coniuge et armis
10.149	*illiso saxo,* qua spina interstruit artus
10.397	*detrito clipeo;* desunt pugnacibus enses
12.180	et *scalis spretis* tentabant rumpere muros
12.523	*tardandis Italis* corruptas igne carinas

$$\alpha^4\,A$$

| 16.379 | *postremo Durio;* pacis de more putares |

$$\alpha^4\,B$$

| 1.637 | vidimus *obsessam patriam* murosque trementes |

$$\alpha^5\,A$$

1.366	*subducti Poeni* vallo caecaque latebra
2.17	*perfusos clipeos* et tela rubentia caede
5.163	*quaesitae fibrae* vanusque moratur haruspex
6.372	*impositus maestus* socium me casibus addo
12.48	intrate atque *epulas promissas* sede Tonantis

14.281	numquam hoste *intratos muros* et quattuor arces
14.307	pascitur *adiutus Vulcanus* turbine venti

$$\alpha^5\,\text{B}$$

7.221	turbari *facilem mentem,* non ultima rerum
16.54	obliquo *telum reflexum* Marte rotabat

$$\alpha^6\,\text{A}$$

2.203	necdum *irae positae;* celsa nam figitur hasta
9.257	*effusae lacrimae,* Mancinique inde reversus
9.364	*laxati cunei,* perque intervalla citatus
15.759	*turbati Rutuli,* confusaque pectora visu

$$\beta^1\,\text{A}$$

6.701	acta *meae dextrae:* captam, Carthago, Saguntum

$$\gamma^1\,\text{A}$$

8.472	lectos Caere *viros, lectos.*Cortona, superbi

$$\gamma^1\,\text{B}$$

2.539	et quidquid *scelerum, poenarum* quidquid et irae

$$\gamma^3\,\text{B}$$

3.302	ad *quorum tactum* mites iacuere cerastae
4.164	*seminecum letum* peragit gravis ungula pulsu
6.403	ecce trahens *geminum natorum* Marcia pignus
12.175	*clarum Volscorum,* Tulli, decus. ast ubi iusso

$$\delta^1\,\text{A}$$

1.145	*occidui qui* solis opes et vulgus Hiberum
1.426	at parte ex *alia, qua* se insperata iuventus
2.363	quis procul *a Tyria* dominos depellere Byrsa
2.490	te matres, vincente *fame, te* maesta virorum
2.532	*hos muros* impelle manu populumque ferocem
3.383	septimaque *his stabulis* longissima ducitur aetas
3.485	*a supera* tellure patet, tam longa per auras
3.675	*has umbras* nemorum et conexa cacumina caelo
3.690	*a prima* venere die; prisco inde pavore
4.406	aspicite. *has dextras* capti genuere parentes
4.420	magnanimi me, nate, *viri, ni* bella capessis
4.670	*di patrii,* quarum auspiciis stat Dardana Roma
4.698	ac tandem *a Trebia* revocavit Scipio fessas

4.809	quid tibi *pro tanto* non impar munere solvat
4.819	vos quoque, *di patrii,* quorum delubra piantur
5.283	aut dumis subit, *albenti si* sensit in aethra
5.476	quae sedit formido, *viri, qui* Marte relicto
7.253	*his dictis* fractus furor, et rabida arma quierunt
7.320	*si cordi* consulta (moras extrema recusant)
7.730	*quis actis* senior numerato milite laetus
8.268	dividit Italiam? an *ferro, quo* cingitis, utor
8.367	nec non *sceptriferi qui* potant Thybridis undam
8.417	Nursia, et *a Tetrica* comitantur rupe cohortes
8.616	armarat *patrio, quo* nutrit bella, metallo
9.172	*his oculis* et damnato violabere visu
9.192	clarior *his titulis* plusque allatura cruoris
9.247	*qua dextra* veniant stridentis sibila belli
9.377	*ni decori* sat sint pariendo tempora leti
10.306	dextera, quem, *soli si* bella agitanda darentur
10.623	*his dictis* sedere minae et conversa repente
11.530	si talis redeat, *populis sis* omnibus una
12.102	hic *pro nubivago* gratus pia templa meatu
12.214	atque Antenorea *sese de* stirpe ferebat
12.701	*his dictis* grates agit ac turbata per auras
13.883	*pro! quanto* levius mortalibus aegra subire
14.219	et *qui praesenti* domitant periura Palici
14.555	crudescunt, ac *pro ferro* stat fluctibus uti
14.665	femineus labor. *his tectis* opibusque potitus
14.684	felices *populi, si,* quondam ut bella solebant
14.686	at, ni cura *viri, qui* nunc dedit otia mundo
15.14	decertandum *hosti, qui* fregerit arma duorum
15.363	*his umeris* tibi opima feram.' nec plura, sereno
15.500	venales animae, *Rhodani qui* gurgite gaudent
15.558	*his dictis* abit atque abscedens visa paventem
15.586	*qui fratri* certet, cui maxima gloria cedat
15.607	castra regi. verum, *fratri si* vita supersit
15.663	quod *ni veloci* prosternimus agmina Marte
16.151	ille *tibi, qui* iam gemino Laurentia lustro
16.275	*his actis* repetit portum puppesque secundo
16.457	*his actis* ductor laeta ad certamina plantae
16.614	litore Elissaeo? *stimuli si* laudis agunt nos

δ¹ B

1.250	imperium proferre *suum. tum* vertice nudo
3.36	ianitor, aeterno *tum primum* tractus ab antro

3.257	Sabratha *tum Tyrium* vulgus Sarranaque Leptis
3.274	*tum primum* castris Phoenicum tendere ritu
4.111	gliscere, et urgebat *trepidam iam* caede priorum
5.52	introitu *facilem, quem* traxit ab aequore, piscem
5.336	pugnas dira manus, *raptum cum* sanguine caesi
5.666	ceu tumulo texere *virum, tum* strage per undas
6.255	et chalybem longo *tum primum* passus in aevo
6.489	talia *quam videam* ferientes pacta Latinos
7.151	*inventum, dum* se cohibet, terimurque secundo
7.723	*tum demum, Tyrium* quas circumfuderat atra
8.131	evasit propere in *celsam, quam* struxerat ante
8.173	*dum caelum* rapida stellas vertigine volvet
8.267	*dum mecum* Garamas et adustus corpore Maurus
9.372	Scaevola; nec tanta *vitam iam* strage volebat
9.650	*iamdudum cum* luce libet. sed comprimit ensem
10.15	ac *iuvenem, quem* Vasco levis, quem spicula densus
10.157	hunc sude, *quam raptam* Libyci per terga iacentis
10.178	*dum laevum* petit incumbens violentius inguen
10.383	sed Varro est victus *quonam tam* prospera Martis
10.431	tum quatiens *strictum cum* voce ante ora paventum
10.489	compressere odia, et *positum cum* foedere bellum
11.155	talia iactantes *iam lectam* sorte parabant
11.158	*tum solum* Decius Capuae decus. isque receptus
11.195	dum Capua *dignum, dum* me duce dextera vindex
11.293	proles digna *deum, tum* Dardanus; isque Tonanti
12.86	arcis templa iugo, *quorum tum* Virrius, altae
12.300	*dum bellum* tonat, et sese furata iuventus
12.572	*tum demum* castris turmas inflexit ovantes
13.168	*tum strictum* propere vagina detegit ensem
13.593	audet adire *ferum, dum* fractis mille catenis
15.491	Pyrenes tumulo *clipeum cum* carmine figunt
15.609	consulibus? sed enim *solum, dum* vera patescant
16.134	Hannibal ipse *etiam iam* iamque modestior armis
17.220	nullaque iam Hesperia et *nusquam iam* Daunia tellus

$$\delta^2 \, A$$

7.273	*huc illuc* castra ac scrutantem proelia Poenum
17.137	*huc illuc* iactans, rectorem prodidit hosti

$$\varepsilon \, A$$

8.97	*si qui* te referant converso flamine venti

ζA

9.424	heu miser! *aequari potuisti* funere Paulo
12.237	*portabis capitis,* quae non sine crimine vester
12.289	vobis terga *dedi; vidi* cum ad bella vocarem
12.347	Hostus erat; *cuius fretus* fulgente iuventa
13.519	sed te *oro, quando* vitae tibi causa labores
13.759	descendisse *Erebo. nullo* non tempus abundans
15.485	*fluminei veluti* deprensus gurgitis undis
16.489	Eurytus et *primus brevibus,* sed primus, abibat
16.685	assueta, et *nobis, quamvis* Oenotria nondum

ζB

7.393	hunc iterum atque *iterum vinctum* vel montibus altis
7.723	*tum demum, Tyrium* quas circumfuderat atra
10.332	hoc *tantum, nondum* portas intrasse Quirini
11.453	namque chaos, *caecam quondam* sine sidere molem
13.467	haud ulla ante *tuam, quamquam* non parva fatigent
14.413	divisitque *fretum, clamantum* spumeus ora
15.94	quippe nec ira *deum tantum* nec tela nec hostes
15.372	iam superis, *magnum secum* portare sub umbras
15.534	culmina *villarum nostrum* delapsa feruntur

STATIUS

	A		B	
	Achilleid (1127 lines)			
γ^1	1	0.09	–	
δ^1	1	0.09	2	0.18
δ^2	1	0.09	–	
ε	1	0.09	–	
ζ	1	0.09	–	
	Silvae (3551 lines)			
α^7	1	0.03	–	
γ^2	1	0.03	–	
γ^3	–		1	0.03
δ^1	16	0.45	2	0.06
δ^2	1	0.03	–	
ε	1	0.03	–	
ζ	4	0.12	2	0.06

	A		B	
Thebaid (9741 lines)				
α^1	2	0.02	1	0.01
α^2	1	0.01	1	0.01
γ^1	3	0.03	–	
δ^1	29	0.30	14	0.14
δ^2	4	0.04	–	
ε	3	0.03	–	
ζ	12	0.13	3	0.03

Very low counts, both alpha and omega, and lowest in his latest work, the
Achilleid.

Achilleid

γ^1 A

1.571   nunc levibus *sertis, lapsis* nunc sponte canistris

δ^1 A

1.222   delphinas *biiugos, quos* illi maxima Tethys

δ^1 B

1.456   conficeret metas, *tum primum* Graecia vires
2.56   concilio *superum, dum* Pelea dulce maritat

δ^2 A

1.200   obtinet, *huc illuc* divisa mente volutat

ε A

1.341   *ne te* suspectum molli non misceat aulae

ζ A

1.407   nec tantum *exciti, bimari* quos Isthmia vallo

Silvae

α^7 A

3.3.133   nec modus aut *pennis laceris* aut crinibus ignem

γ^2 A

3.5.45   qua veteres *Latias Graias* heroidas aequas

$$\gamma^3 B$$

1.3.39	*lucorum senium?* te, quae vada fluminis infra

$$\delta^1 A$$

1.2.250	quisque lyra. sed *praecipui, qui* nobile gressu
1.4.31	quare age, *si Cereri* sua dona merumque Lyaeo
1.4.55	imperium vigilesque *suo pro* Caesare curae
1.4.58	tunc deus *Alpini qui* iuxta culmina dorsi
2.1.29	*si merui,* luctusque tui consortia sensi
2.1.215	ut sileam *morbos; hos* ora rigentia brumae
2.5.18	*a media* iam morte redit, nec protinus omnes
3.4.101	*si merui,* longa dominum renovare iuventa
4.1.21	continua: *hos umeros* multo sinus ambiat ostro
4.4.56	at tu, *si longi* cursum dabit Atropos aevi
4.8.45	*di patrii,* quos auguriis super aequora magnis
5.2.152	felix *qui viridi* fidens, Optate, iuventa
5.3.36	ordior acclinis *tumulo, quo* molle quiescis
5.3.179	*qui Diomedei* celat penetralia furti
5.5.55	dulce *sibi, si* busta Lini complexus Apollo
5.5.60	imperii, Fortuna, *tui, qui* dicere legem

$$\delta^1 B$$

1.2.47	attulit? hic *mecum, dum* fervent agmine postes
1.2.139	saepe neget maerens *ipsam iam* cedere sensi

$$\delta^2 A$$

1.3.72	*huc illuc* fragili prosternit pectora musco

$$\varepsilon A$$

5.3.126	*te de* gente suum Latiis ascita colonis

$$\zeta A$$

1.2.76	*quamvis Ausoniis* multum gener ille petitus
2.5.27	ingemuere *mori; magni* quod Caesaris ira
2.6.13	ne *pudeat; rumpat* frenos dolor iste diesque
3.5.35	aure rapis *vigili; longi* tu sola laboris

$$\zeta B$$

3.3.77	*nondum stelligerum* senior dimissus in axem
5.3.75	parcius *exstinctum saevorum* crimine agrestum

Thebaid

α^1 A

| 1.489 | Tydea per *latos umeros* ambire laborant |
| 2.542 | terga super *laevos umeros* vicina cruori |

α^1 B

| 3.541 | *maiorum volucrum* tenerae deponitis alae |

α^2 A

| 6.250 | fama vocat; *cunctis arvis* ac moenibus adsunt |

α^2 B

| 4.152 | dat tamen haec *iuvenum tercentum* pectora; vulgus |

γ^1 A

11.680	iam tumulis *victos, socios* iam moenibus arces
12.513	*deveniunt, cedunt* miserorum turba priorum
12.627	hos Salamin *populos; illos* Cerealis Eleusin

δ^1 A

1.60	si bene quid *merui, si me de* matre cadentem
1.62	*firmasti, si* stagna peti Cirrhaea bicorni
1.73	*exaudi, si* digna precor quaeque ipsa furenti
1.114	ut stetit, *abrupta qua* plurimus arce Cithaeron
1.271	incipe, *fluctivaga qua* praeterlabitur unda
2.85	luxus: at *Ogygii si* quando afflavit Iacchi
2.245	*Inachidae, quae* cuique domus sacrique facultas
4.367	lumina, *si duri* comperta clade sequantur
5.280	litus eunt: *illa, qua* rere silentia, porta
5.324	immeritasque manus; *subeo—pro* dira potestas!—
5.629	ferte, duces, *meriti si* qua est mihi gratia duri
5.683	pergite in excidium, *socii si* tanta voluptas
6.341	Hypsipyles, subiere *iugo, quo* vectus uterque
6.414	evolvere globum et *spatio quo* quisque valebat
6.515	*te Thebe* fraterque palam, te plangeret Argos
6.516	*te Nemee,* tibi Lerna comas Larissaque supplex
7.521	et *lacrimas has* quisque domi: sua credite matri
7.539	me potius, *socii, qui* fidum Eteoclea nuper
7.707	qui tripodas laurusque *sequi, qui* doctus in omni
8.742	i, precor, *Atrei si* quid tibi sanguinis umquam
9.197	auxilium et seram *rapto pro* corpore pugnam

9.623	hunc, precor, *audaci qui* nunc ad proelia voto
10.419	pectora, et *a media* catulos circumspicit ira
10.449	at ferus Amphion, *regi qui* facta reportent
10.920	mirantur taciti et *dubio pro* fulmine pallent
12.246	haud procul *exacti si* spes non blanda laboris
12.264	redde mihi, tuque, oro, *veni, si* manibus ulla
12.267	*si merui.'* dixit, tectumque aggressa propinquae
12.615	qui gelidum Braurona *viri, qui* rura lacessunt

δ¹ B

1.350	*dum caelum* sibi quisque rapit; sed plurimus Auster
1.500	indulgens reparare *animum, dum* proximus aegris
4.783	*dum parvum* circa iubet exsultare Tonantem
5.291	Lucifer astra polo. *tum demum* litore rauco
5.429	dant Fluvii Montesque *locum, tum* Terra superbit
5.472	efferus, o *utinam iam* tum mea litora rectis
6.475	ausa venit. ruit, *Haemonium dum* fervidus instat
8.13	augure *tum demum* rumpebant stamina Parcae
8.638	fertur Atys, servans *animam iam* sanguine nullo
9.116	infestante lupo *tum primum* feta tuetur
9.196	at ferus Hippomedon *quamquam iam* sentit inane
11.140	viderat, obscura *vallum dum* nocte pererrat
12.351	moenibus egressus: *illam nam* tempore in omni
12.443	vicit nempe Creon. *nusquam iam* regna, quis ardor

δ² A

2.602	*huc illuc* clipeum obiectans, seque ipse recedens
4.366	decedit stabulis *huc illuc* turbida verrens
4.733	flammatum pecus. *huc illuc* impellit Adrastus
9.172	*huc illuc* frustra ruit avius, impia donec

ε A

1.60	si bene quid *merui, si me de* matre cadentem
3.274	hoc mihi Lemniacae *de te* meruere catenae
9.662	nec *te de* dubiis fraterna oracula fallunt

ζ A

2.644	protinus idem *ultro iaculo* parmaque Menoeten
3.169	nec minus *interea digesta* strage suorum
3.342	regis *Echionii: fratri* sua iura negari
4.80	*conciliat; dederat* nec non socer ipse regendas
4.544	Thebanasque *animas; alias* avertere gressus

6.145   Cadme, *domus, nullus* Tyrio grege plangitur infans
6.854   ille autem, *Alpini veluti* regina cupressus
7.47   digna *loco statio:* primis salit Impetus amens
8.101   crimine non *ullo subeo* nova fata nec alma
9.141   ter retrahunt *Danai: Siculi* velut anxia puppis
9.441   Liber? an *Eous melius* pacatur Hydaspes
10.165   impatiensque *dei, fragili* quem mente receptum

ζ B

2.372   *Aetolum, multum* lacrimis conata morari
4.247   a rudis *armorum, tantum* nova gloria suadet
10.569   attoniti et *tantum matrum* lamenta trementes

TIBULLUS (1240 lines)

	A		B	
α^2	1	0.08	–	
α^6	1	0.08	–	
δ^1	7	0.57	5	0.40
δ^2	1	0.08	–	
ε	3	0.25	–	
ζ	2	0.16	1	0.08

α^2 A

2.4.60   mille *alias herbas* misceat illa, bibam

α^6 A

2.3.14 a   ipse *deus solitus* stabulis expellere vaccas

δ^1 A

1.2.15   tu quoque *ne timide* custodes, Delia, falle
1.2.17   illa favet, seu *quis iuvenis* nova limina temptat
1.2.85   non ego, *si merui,* dubitem procumbere templis
1.7.61   te canet agricola, *a magna* cum venerit urbe
1.9.34   non tibi *si Bacchi* cura Falernus ager
2.1.17   *di patrii,* purgamus agros, purgamus agrestes
2.6.36   illius ut *verbis, sis* mihi lenta veto

$$\delta^1\,B$$

1.4.66	*dum caelum* stellas, dum vehet amnis aquas
1.5.71	non frustra *quidam iam* nunc in limine perstat
1.8.43	*tum studium* formae est: coma tum mutatur, ut annos
2.1.85	aut etiam sibi quisque *palam: iam* turba iocosa
2.3.5	o ego, dum aspicerem *dominam, quam* fortiter illic

$$\delta^2\,A$$

1.3.70	saevit, et *huc illuc* impia turba fugit

$$\varepsilon\,A$$

1.4.15	sed *te ne* capiant, primo si forte negarit
1.6.8	sic etiam *de me* pernegat usque viro
1.9.72	devoveat *pro quo* teque domumque tuam

$$\zeta\,A$$

1.3.4	*abstineas avidas,* Mors, precor, atra, manus
2.5.110	et *faveo morbo* quem iuvat ipse dolor

$$\zeta\,B$$

2.4.22	ne *iaceam clausam* flebilis ante domum

VALERIUS FLACCUS (5591 lines)

	A		B	
β^2	1	0.02	–	
γ^1	2	0.04	–	
δ^1	21	0.38	15	0.27
δ^2	1	0.02	–	
ε	3	0.05	2	0.04
ζ	5	0.09	1	0.02

W.-W. Ehlers' note on 1.342, '-$\bar{o}$-$\bar{o}$ susp.', illustrates prevailing ignorance. Valerius' total abstinence from alphas and low count of omegas does not preclude such homs as this. Likewise the supplement in 3.298 *meque tuus* ⟨*potius*⟩ *nunc plangeret error* (or *melius*) is unobjectionable, though *fusum* would do as well.

The suspicion (Håkanson, 91) that 'further investigation might show that homoeoteleuta are more frequent toward the end of this unfinished epos' has proved unwarranted, as also in Lucan's case. Håkanson quotes from Housman's preface (p. XXX): 'To protect faults from amendment, and shift the blame of them from the scribes to the poet, it is sometimes urged that this poem is unfinished, and that books IV–X, not published by Lucan himself, must be expected to contain blemishes due to lack of revision. Such blemishes should be distinguishable by their nature, and in the last two books I do seem to find them ... but not elsewhere.'

$$\beta^2 A$$

7.245   quaere *malis nostris* requiem mentemque repone

$$\gamma^1 A$$

3.245   *explorant, prensant* socios vocemque reposcunt
8.408   sed miser ut *vanos, veros* ita saepe timores

$$\delta^1 A$$

1.285   coeperat *a gemina* discedere Sestos Abydo
1.652   litora depellitque *Notos, quos* caerulus horror
1.676   *has animas* patriaeque amplecti limina portae
2.96   aut Lipares domus. *has epulas,* haec templa peracta
2.380   redde 'ait' *Aesonide. me* tecum solus in aequor
3.159   Dorcea, *qui dulci* festis assistere mensis
3.324   solus et *a prima* fueras spes una iuventa
3.336   *funereae*[27] *quae* cuique manus. quae cura suorum
4.202   signa manent; instar *scopuli, qui* montibus altis
4.321   occupat, *annosi veluti si* decidat olim
4.365   aut *quos inventos*[28] tenuisset Iuppiter astus
4.398   qualis et *a prima* quantum mutata iuventa
5.19   *si nostri* te cura movet, qui cardine summo
5.52   nunc quoque, *si tenui* superant in imagine curae
5.238   praeterea *infernae quae* nunc sacrata Dianae
5.378   si dea, *si magni* decus huc ades, inquit, Olympi
6.100   *hiberni qui* terga Novae gelidumque securi
6.311   et tibi si pietas *nati, si* dextra fuisset
7.501   umquam ego *si meriti* sim noctis et immemor huius
8.242   cogitet aut *ipsi qui* iam meminere ministri
8.280   *qui potui:* patriae veniam da, quaeso, senectae
8.442   *Aesonide, me* ferre preces et supplicis ora

[27] Not counted. *munera fert* Langen.
[28] Conjectural reading.

δ¹ B

1.85	ante *deam quam* te tonitru nutuque reposci
1.612	*nimborum cum* prole Notus crinemque procellis
1.626	ignari, sed tale *fretum. tum* murmure maesto
1.660	Aesoniamque capit *pateram, quam* munere gaudens
1.847	inducitque *nurum, tum* porta quanta sinistra
2.445	Thessala Dardaniis *tum primum* puppis harenis
3.183	*tum primum* puer ausus Hylas (spes maxima bellis
3.633	saltibus ut mediis *tum demum* laeta reducit
3.680	tendite, *dum rerum* patiens dolor et rude membris
4.33	impulit Oenides. *verum cum* gente domoque
5.273	iamque aderat magnis *regum cum* milibus urbi
6.188	hinc gemitus mixtaeque *virum cum* pulvere vitae
6.250	*nequiquam. iam* cuspis inest nec fragmina curat
6.396	aegida *tum primum* virgo spiramque Medusae
8.214	ipse dab⟨at, i⟩am *nubiferam* transire Carambin

δ² A

4.257	*hinc illinc* dubiis intenta silentia votis

ε A

1.332	nec *de te* credam nostris ingrata serenis
3.317	cuncta trahis. necdum suboles nec gaudia *de te*
5.404	ventum, ait, huc adytis iam *se de* more paternis

ε B

2.154	exanimat, *quam iam* miseros transversa tuentem
4.697	aut sociis temptata quies, nigrantia *quam iam*

ζ A

1.342	vota deis: *video nostro* tot in aequore reges
4.321	occupat, *annosi veluti si* decidat olim
5.223	Soligenae *falli meriti* meritique relinqui
5.261	signa *mali; reddi* iubet exitiale sacerdos
8.344	hoc adeo *interea specta* de litore pugnas

ζ B

1.340	concussoque *ratem gauderem* tollere remo

VIRGIL

	A		B	
	Aeneid (9891 lines)			
α^1	25	0.26	16	0.16
α^2	4	0.04	–	
α^3	3	0.03	–	
α^4	1	0.01	–	
α^5	2	0.02	–	
α^6	1	0.01	–	–
α^7	1	0.01	–	
β^1	2	0.02	–	
γ^2	–		1	0.01
δ^1	52	0.53	24	0.24
δ^2	3	0.03	–	
ε	13	0.13	–	
ζ	5	0.05	3	0.03

	A		B	
	Eclogues (829 lines)			
α^1	1	0.12	–	
α^4	1	0.12	–	
β^1	2	0.24	–	
β^2	1	0.12	–	
γ^1	1	0.12	–	
δ^1	6	0.72	–	
ε	2	0.24	–	
ζ	1	0.12	–	
	Georgics (2186 lines)			
α^1	1	0.05	1	0.05
α^3	–		1	0.05
γ^2	2	0.09	–	
γ^3	–		1	0.05
δ^1	6	0.27	5	0.23
δ^2	1	0.05	–	

'One may ask whether the considerable difference between the *Georgics* and the *Aeneid* is to some extent due to the fact that the *Aeneid* was never submitted to the poet's *secunda cura*' (Håkanson, 93). It can now be appreciated that the difference is far more than considerable and that the answer to the question is an unhesitant no. The percentage of α^1s in the later work is about four times as high as in the two earlier. Moreover, the solitary α^1 in the *Eclogues*, 6.55 f. *Nymphae,/Dictaeae Nymphae,* is evidently a special effect (cf. *Aen.* 8.71 *Nymphae, Laurentes Nymphae*), and one of the two in the *Georgics* occurs in a verse (3.219) whose unusual rhythm and word order is remarked by commentators. Surely the poet felt, even if unconsciously, that the 'neoteric' practice did not befit his epic, to which he wished to lend an Ennian flavor. That some of its α^1s may actually come from Ennius is, of course, quite likely.

Aeneid

α^1 A

1.225	litoraque et *latos populos,* sic vertice caeli
2.251	involvens *umbra magna* terramque polumque
2.721	haec fatus *latos umeros* subiectaque colla

3.282   *nudati socii;* iuvat evasisse tot urbes
3.536   *turriti scopuli,* refugitque ab litore templum
4.255   *piscosos scopulos* humilis volat aequora iuxta
5.190   *Hectorii socii,* Troiae quos sorte suprema
5.376   ostenditque *umeros latos* alternaque iactat
6.93   causa *mali tanti* coniunx iterum hospita Teucris
6.638   devenere *locos laetos* et amoena virecta
6.680   *inclusas animas* superumque ad lumen ituras
6.887   aeris in *campis latis* atque omnia lustrant
7.401   si qua *piis animis* manet infelicis Amatae
7.776   solus ubi in *silvis Italis* ignobilis aevum
8.261   *elisos oculos* et siccum sanguine guttur
9.256   Ascanius, *meriti tanti* non immemor umquam
9.725   obnixus *latis umeris,* multosque suorum
10.335   *Iliacis campis.'* tum magnam corripit hastam
10.510   nec iam fama *mali tanti,* sed certior auctor
11.24   ite, ait, *egregias animas,* quae sanguine nobis
11.355   quin natam *egregio genero* dignisque hymenaeis
11.480   causa *mali tanti,* oculos deiecta decoros
11.679   cui pellis *latos umeros* erepta iuvenco
12.40   quid *consanguinei Rutuli,* quid cetera dicet
12.100   vibratos *calido ferro* murraque madentis

α¹ B

1.569   seu vos *Hesperiam magnam* Saturniaque arva
3.349   procedo et *parvam Troiam* simulataque magnis
4.143   qualis ubi *hibernam Lyciam* Xanthique fluenta
4.345   sed nunc *Italiam magnam* Gryneus Apollo
5.311   alter *Amazoniam pharetram* plenamque sagittis
5.682   stuppa vomens *tardum fumum,* lentusque carinas
6.179   itur in *antiquam silvam,* stabula alta ferarum
6.270   quale per *incertam lunam* sub luce maligna
6.634   corripiunt *spatium medium* foribusque propinquant
6.639   *fortunatorum nemorum* sedesque beatas
6.812   misus in *imperium magnum.* cui deinde subibit
8.389   accepit *solitam flammam,* notusque medullas
9.403   suspiciens *altam Lunam*[29] sic voce precatur
11.234   ergo *concilium magnum* primosque suorum
11.545   *solorum nemorum;* tela undique saeva premebant
12.566   neu quis ob *inceptum subitum* mihi segnior ito

---

[29] Reading doubtful.

$$\alpha^2 \, A$$

2.503	quinquaginta *illi thalami,* spes tanta nepotum
6.411	inde *alias animas,* quae per iuga longa sedebant
9.782	*quos alios muros,* quae iam ultra moenia habetis
11.352	unum etiam *donis istis,* quae plurima mitti

$$\alpha^3 \, A$$

3.404	et *positis aris* iam vota in litore solves
5.271	*amissis remis* atque ordine debilis uno
12.129	utque *dato signo* spatia in sua quisque recessit

$$\alpha^4 \, A$$

4.321	odere, *infensi Tyrii.* te propter eundem

$$\alpha^5 \, A$$

1.482	diva solo *fixos oculos* aversa tenebat
6.469	illa solo *fixos oculos* aversa tenebat

$$\alpha^6 \, A$$

12.269	*turbati cunei* calefactaque corda tumultu

$$\alpha^7 \, A$$

10.450	aut leto *insigni; sorti* pater aequus utrique est

$$\beta^1 \, A$$

6.67	regna *meis fatis)* Latio considere Teucros
9.543	confixique *suis telis* et pectora duro

$$\gamma^2 \, B$$

9.270	aureus; *ipsum illum, clipeum* cristasque rubentis

$$\delta^1 \, A$$

1.201	accestis *scopulos, vos* et Cyclopia saxa
1.418	corripuere viam *interea, qua* semita monstrat
1.606	saecula? *qui tanti* talem genuere parentes
1.753	immo age et *a prima* dic, hospes, origine nobis
2.183	hanc *pro Palladio* moniti, pro numine laeso
2.282	*quae tantae* tenuere morae? quibus, Hector, ab oris
2.379	improvisum aspris *veluti qui* sentibus anguem
2.538	debita, *qui nati* coram me cernere letum
2.702	*di patrii,* servate domum, servate nepotem
3.12	cum sociis natoque Penatibus et *magnis dis*

3.149	quos mecum *a Troia* mediisque ex ignibus urbis
4.54	*his dictis* incensum animum inflammavit amore
4.314	mene fugis? per ego *has lacrimas* dextramque tuam te
4.419	hunc ego *si potui* tantum sperare dolorem
5.347	ultima, *si primi* Salio reddentur honores
5.355	digna dabis, primam *merui qui* laude coronam
5.357	et simul *his dictis* faciem ostentabat et udo
5.544	extremus *volucri qui* fixit harundine malum
6.111	eripui *his umeris* medioque ex hoste recepi
6.133	quod si tantus amor *menti, si* tanta cupido est
6.236	*his actis* propere exsequitur praecepta Sibyllae
6.318	'dic' ait 'o virgo, quid vult concursus ad amnem
6.335	quos simul *a Troia* ventosa per aequora vectos
6.382	*his dictis* curae emotae pulsusque parumper
6.666	*quos circumfusos* sic est affata Sibylla
7.98	externi venient *generi, qui* sanguine nostrum
7.263	ipse modo Aeneas, *nostri si* tanta cupido est
8.15	quid struat *his coeptis,* quem, si Fortuna sequatur
8.51	Arcades *his oris,* genus a Pallante profectum
8.538	*quas poenas* mihi, Turne, dabis! quam multa sub undas
8.679	cum patribus populoque, Penatibus et *magnis dis*
8.715	moenia, *dis Italis* votum immortale sacrabant
9.238	qui patet in bivio *portae quae* proxima ponto
9.247	*di patrii,* quorum semper sub numine Troia est
9.782	*quos alios muros,* quae iam ultra moenia habetis
10.362	at parte ex *alia, qua* saxa rotantia late
10.371	spemque meam, *patriae quae* nunc subit aemula laudi
10.440	Turnum, *qui volucri* curru medium secat agmen
11.54	*hi nostri* reditus exspectatique triumphi
11.73	extulit *Aeneas, quas* illi laeta laborum
11.361	proicis, *o Latio* caput horum et causa malorum
11.417	egregiusque *animi, qui,* ne quid tale videret
11.576	pro crinali *auro, pro* longae tegmine pallae
11.827	iamque vale.' simul *his dictis* linquebat habenas
12.152	tu *pro germano* si quid praesentius audes
12.233	vix hostem, *alterni si* congrediamur, habemus
12.245	dat signum *caelo, quo* non praesentius ullum
12.351	illum Tydides *alio pro* talibus ausis
12.627	sunt *alii, qui* tecta manu defendere possint
12.658	*quos generos* vocet aut quae sese ad foedera flectat
12.749	inclusum *veluti si* quando flumine nactus
12.843	*his actis* aliud genitor secum ipse volutat

δ¹ B

1.193	corpora fundat humi et *numerum cum* navibus aequet
1.635	terga suum, pinguis *centum cum* matribus agnos
1.729	implevitque mero *pateram, quam* Belus et omnes
2.559	at me *tum primum* saevus circumstetit horror
3.177	ad *caelum cum* voce manus et munera libo
3.381	principio *Italiam, quam* tu iam rere propinquam
4.371	quae quibus *anteferam? iam iam* nec maxima Iuno
4.421	exsequere, Anna, mihi: *solam nam* perfidus ille
5.199	subtrahiturque *solum; tum* creber anhelitus artus
5.260	*loricam, quam* Demoleo detraxerat ipse
7.294	fata *Phrygum! num* Sigeis occumbere campis
7.762	Virbius, *insignem quem* mater Aricia misit
8.222	*tum primum* nostri Cacum videre timentem
8.342	hinc lucum *ingentem, quem* Romulus acer asylum
9.285	est mihi, *quam miseram* tenuit non Ilia tellus
9.590	*tum primum* bello celerem intendisse sagittam
9.815	*tum demum* praeceps saltu sese omnibus armis
10.58	*dum Latium* Teucri recidivaque Pergama quaerunt
10.589	fulgentis clipei, *tum laevum* perforat inguen
11.113	nec *bellum cum* gente gero: rex nostra reliquit
11.739	(hic amor, hoc *studium*) *dum* sacra secundus haruspex
12.6	*tum demum* movet arma leo, gaudetque comantis
12.90	*ensem, quem* Dauno ignipotens deus ipse parenti
12.213	conspectu in medio *procerum. tum* rite sacratas

δ² A

4.363	*huc illuc* volvens oculos, totumque pererrat
5.408	*huc illuc* vinclorum immensa volumina versat
12.764	*huc illuc;* neque enim levia aut ludicra petuntur

ε A

3.604	*pro quo,* si sceleris tanta est iniuria nostri
4.317	si bene quid *de te* merui fuit aut tibi quicquam
4.327	saltem si qua mihi *de te* suscepta fuisset
6.189	heu nimium *de te* vates, Misene, locuta est
6.502	cui tantum *de te* licuit? mihi fama suprema
9.207	Nisus ad haec: 'equidem *de te* nil tale verebar
9.491	et funus lacerum tellus habet? hoc mihi *de te*
10.743	nunc morere, ast *de me* divum pater atque hominum rex
11.45	non haec Evandro *de te* promissa parenti

11.302	ante quidem summa *de re* statuisse, Latini
12.72	ne, quaeso, *ne me* lacrimis neve omine tanto
12.801	*ne te* tantus edit tacitum dolor et mihi curae
12.875	iam iam linquo acies. *ne me* terrete timentem

ζ A

1.737	primaque *libato summo* tenus attigit ore
3.110	Pergameae *steterant; habitabant* vallibus imis
3.630	nam simul *expletus dapibus* vinoque sepultus
5.306	Cnosia bina *dabo levato* lucida ferro
8.144	temptamenta *tui pepigi;* me, me ipse meumque

ζ B

3.665	iam *medium, necdum* fluctus latera ardua tinxit
6.634	corripiunt *spatium medium* foribusque propinquant
9.101	mortalem *eripiam formam* magnique iubebo

Eclogues

α^1 A

6.56	*Dictaeae Nymphae,* nemorum iam claudite saltus

α^4 A

10.39	et *nigrae violae* sunt et vacinia nigra

β^1 A

7.22	quale *meo Codro,* concedite; proxima Phoebi
10.2	pauca *meo Gallo,* sed quae legat ipsa Lycoris

β^2 A

5.18	*iudicio nostro* tantum tibi cedit Amyntas

γ^1 A

3.88	qui te, Pollio, *amat, veniat* quo te quoque gaudet

δ^1 A

4.8	tu modo nascenti *puero, quo* ferrea primo
5.38	pro molli viola, *pro purpureo narcisso*
8.40	iam fragilis poteram *a terra* contingere ramos
8.95	*has herbas* atque haec Ponto mihi lecta venena
9.28	Mantua, *vae, miserae* nimium vicina Cremonae
10.46	tu procul *a patria* (nec sit mihi credere tantum)

εA

3.4   dum fovet, ac, *ne me* sibi praeferat illa veretur
10.48   me sine sola vides. a, *te ne* frigora laedant

ζA

8.75   effigiem *duco; numero* deus impare gaudet

Georgics

α¹A

3.219   pascitur in *magna Sila* formosa iuvenca

α¹B

3.520   non umbrae *altorum nemorum,* non mollia possunt

α³B

2.399   terque quaterque *solum scindendum,* glaebaque versis

γ²A

1.138   *Pleiadas, Hyadas* claramque Lycaonis Arcton
2.169   extulit, haec *Decios, Marios* magnosque Camillos

γ³B

1.474   *armorum sonitum* toto Germania caelo

δ¹A

1.354   *quo signo* caderent austri; quid saepe videntes
1.405   et *pro purpureo* poenas dat Scylla capillo
2.259   *his animadversis* terram multo ante memento
3.123   *his animadversis* instant sub tempore et omnis
4.537   sed modus *orandi qui* sit, prius ordine dicam
4.565   carmina *qui lusi* pastorum, audaxque iuventa

δ¹B

3.205   *tum demum* crassa magnum farragine corpus
3.325   carpamus, dum mane *novum, dum* gramina canent
3.463   et lac *concretum cum* sanguine potat equino
4.110   et custos furum atque *avium cum* falce saligna
4.490   restitit, Eurydicenque *suam iam* luce sub ipsa

δ²A

2.297   *huc illuc,* media ipsa ingentem sustinet umbram

LATE ANTIQUITY

	α^1	α^{1-6}	ω
Alcimus Avitus (3218)	0.12 (0.09 + 0.03)	0.12 (0.09 + 0.03)	1.51 (0.87 + 0.64)
* Ausonius (3267)[1]	0.12 (0.06 + 0.06)	0.18 (0.09 + 0.09)	1.07 (0.73 + 0.34)
* Avianus (654)	0.15 (0.15 + 0)	0.15 (0.15 + 0)	0.92 (0.46 + 0.46)
* Avienius			
Aratea (1878)	0.16 (0.16 + 0)	0.31 (0.31 + 0)	1.58 (1.38 + 0.20)
Descriptio (1393)	0.14 (0.14 + 0)	0.14 (0.14 + 0)	1.44 (1.08 + 0.36)
* Claudian (8333)	0.16 (0.10 + 0.06)	0.22 (0.16 + 0.06)	1.20 (0.85 + 0.35)
Corippus			
Iohannis (4664)	0.24 (0.24 + 0)	0.32 (0.28 + 0.04)	1.80 (1.29 + 0.51)
Laudes Iustini (1680)	1.60 (1.44 + 0.16)	1.97 (1.73 + 0.24)	2.29 (1.79 + 0.50)
Cyprianus Gallus (1498)	0.34 (0.07 + 0.27)	0.61 (0.27 + 0.34)	2.33 (1.40 + 0.93)
* Disticha Catonis (290)	0.34 (0.34 + 0)	0.34 (0.34 + 0)	3.10 (2.76 + 0.34)
Dracontius			
De laudibus Dei (2327)	0.30 (0.26 + 0.04)	0.42 (0.34 + 0.08)	2.18 (1.45 + 0.73)
* *Orestes* (974)	0.30 (0.30 + 0)	0.30 (0.30 + 0)	1.82 (1.02 + 0.80)
* *Romulea* (2354)	0.08 (0.04 + 0.04)	0.20 (0.12 + 0.08)	1.10 (0.85 + 0.25)
Satisfactio (316)	–	–	1.85 (1.54 + 0.31)
Ennodius (1500)	0.53 (0.33 + 0.20)	0.60 (0.40 + 0.20)	0.86 (0.53 + 0.33)
* Epistula Didonis (150)	–	–	0.67 (0 + 0.67)
Juvencus (3211)	1.27 (0.62 + 0.65)	1.68 (0.93 + 0.75)	2.18 (1.43 + 0.75)
* Laus Herculis (137)	–	–	2.92 (0.73 + 2.19)
Marius Victor (2020)	0.15 (0.10 + 0.05)	0.35 (0.30 + 0.05)	1.79 (1.29 + 0.50)
* Maximianus (686)	0.44 (0.44 + 0)	0.73 (0.58 + 0.15)	2.04 (1.02 + 1.02)
* Merobaudes (271)	–	–	0.37 (0 + 0.37)
* Nemesianus			
Cynegetica (325)	0.31 (0.31 + 0)	0.31 (0.31 + 0)	1.23 (0.31 + 0.92)
Eclogues (250)	–	–	4.40 (2.40 + 2.00)
Orientius (1036)	0.29 (0.29 + 0)	0.58 (0.58 + 0)	1.73 (1.25 + 0.48)
* Palladius (170)	0.59 (0.59 + 0)	0.59 (0.59 + 0)	0.59 (0.59 + 0)
Paulinus of Nola (5349)	0.39 (0.30 + 0.09)	0.52 (0.41 + 0.11)	1.39 (0.84 + 0.55)
Paulinus of Pella (616)	0.98 (0.49 + 0.49)	1.14 (0.65 + 0.49)	1.95 (1.30 + 0.65)
Paulinus of Petricordia (3622)	0.25 (0.17 + 0.08)	0.33 (0.21 + 0.11)	1.96 (0.86 + 1.10)
* Phoenix (170)	1.18 (0.59 + 0.59)	1.18 (0.59 + 0.59)	0.59 (0.59 + 0)
* Piscatoris furtum (285)	0.35 (0.35 + 0)	0.70 (0.35 + 0.35)	0.70 (0 + 0.70)
* Priscianus			
Laud. Anast. (312)	0.96 (0.96 + 0)	1.25 (1.25 + 0)	2.56 (2.24 + 0.32)
Periegesis (1087)	0.47 (0.28 + 0.19)	1.02 (0.83 + 0.19)	2.76 (1.56 + 1.20)

[1] Seculars are asterisked.

Prudentius (3013)	0.74 (0.37 + 0.37)	0.83 (0.37 + 0.46)	1.46 (0.66 + 0.80)
* Reposianus (182)	–	–	2.20 (2.20 + 0)
* Rutilius (712)	0.14 (0.14 + 0)	0.28 (0.14 + 0.14)	1.68 (1.40 + 0.28)
Sedulius (1753)	0.06 (0.06 + 0)	0.11 (0.11 + 0)	1.60 (1.03 + 0.57)
* Serenus (1107)	0.09 (0.09 + 0)	0.09 (0.09 + 0)	0.54 (0.45 + 0.09)
* Sidonius (1847)	0.44 (0.33 + 0.11)	0.65 (0.49 + 0.16)	2.10 (1.13 + 0.97)
* Symposius (317)	2.51 (2.20 + 0.31)	3.43 (3.12 + 0.31)	2.50 (1.90 + 0.60)

The Latin verse of late antiquity divides easily into secular and religious
(Christian), the former more or less closely approaching classical standards
of technique, the latter usually uncouth and wooden. The distinction is in
the main reflected in the incidence of alpha homs.

In 25752 secular hexameters or pentameters accounted for on my list[2] I
find 56 α^1s and 23 α^{2-6}s, yielding percentages of 0.22 and 0.09. In 29469
religious I find 130 and 42: percentages 0.44 and 0.14. But as with the
classical poets, both groups have their noncomformists. Maximian and Sidonius are relatively tolerant of alphas, Symposius, generally accounted a
good versifier, wallows in them. Eliminate these three, and the secular percentages drop to 0.14 and 0.05, high by classical standards, but lower than
Silius's and close to those of the later (non-ovidian?) *Heroides* and Juvenal.
Among the religious, Sedulius and Alcimus Avitus show themselves decid-
edly allergic to alphas.[3] Without them the percentages rise to 0.57 and 0.16.
The overall omega count also stands higher among the religious, though not
by a great deal: 1.70 against 1.40.[4]

ALCIMUS AVITUS

Poemata (3218 lines)

	A		B	
α^1	3	0.09	1	0.03
γ^1	3	0.09	–	
γ^2	3	0.09	–	
γ^3	–		4	0.12
δ^1	21	0.65	14	0.43
ε	1	0.03	–	
ζ	7	0.22	1	0.03

[2] I have not troubled to separate Merobaudes' 30 lines *De Christo* from his otherwise
secular survivals. Corippus and Priscianus are set aside, the former as falling somewhere
in between and both for their late date.

[3] So too Marius Victor so far as α^1s are concerned.

[4] Approximate figures.

α¹ A

3.271	o pater, *electas animas* qui sede beata
3.288	*praeteritae vitae* sortem deponis; uterque
5.527	*Hebraeus populus* rubranti castra profundo

α¹ B

5.301	*iustitium iustum* cogit maerere merentes

γ¹ A

2.302	quod removere *nequit, duplicavit* iustius ira
5.209	*continuit, vidit* nullum nec visus ab ullo est
6.337	caeditur hic *multis; paucis* qui nescius errat

γ² A

1.84	olfactu *auditu visu* gustuque potentem
2.31	cessabit *gemitus luxus* metus ira voluptas
6.435	ira *furor maeror livor* discordia luxus

γ³ B

1.275	inque *locum pecorum* viridantis iugere campi
2.35	his *protoplastorum sensum* primordia sacra
4.124	*consensum scelerum* turbata superbia rupit
5.138	*lympharum damnum* proprio supplere cruore

δ¹ A

1.191	*pro thalamo* paradisus erat, mundusque dabatur
1.229	arboribusque *comae: quae,* cum se flore frequenti
2.6	at *si curvati* fecundo pondere rami
2.396	succedens *homini, si* non sal fauce notetur
3.44	caelorum clangente *tuba, qua* nuntius ante
3.165	nam *pro triticeo* lolium consurgere fructu
3.350	ius anceps pugnare *foro, quo* iurgia fratrum
4.98	supplerent vasti *mixto pro* crure dracones
4.382	ipse etiam dignus *tali qui* tempore princeps
4.467	illustres fluvii, *magnos quos* inclita cursu
5.37	*quo signo* spinas nostrae fervescere mentis
5.68	*quo baculo* nitens *gressum tum* dextra regebat
5.181	praeterit ille dies: *tum demum* luce secuta
5.331	*his dictis* lacrimas populus dedit. ipse superbus
5.516	spectet, *quos laetos* vix possit cernere, vultus
5.550	*digni qui* tantos nequeant sentire dolores

6.217	defuerit *facti, si* Christum credula corde
6.225	gustabat sumptam *nostro pro* crimine mortem
6.345	mirantes hortata *viros, quos* ipsa ducatu
6.369	*pro gladio* semper verbum teneatur acutum
6.596	cum sumat tamen iste *cibos, quos* vertice pendens

$$\delta^1 B$$

1.51	*tum demum* tali sapientia voce locuta est
2.75	sic natura valet, *rectam quam* condidit auctor
2.253	intereunte anima *letum dum* crimina poscunt
2.273	*tum primum* nudos (dubium, quid dicere possim)
2.292	*iam magicam* digne valeat quis dicere fraudem
3.99	*quam sociam* misero prima sub lege dedisti
3.261	nec eius *similem, quem* dudum luce receptum
3.321	excitat ad pugnam *tum primum* conscia virtus
3.327	*tum primum* tectis taetra caligine caedis
3.415	sicque reus *scelerum, dum* digna piacula pendit
5.68	*quo baculo* nitens *gressum tum* dextra regebat
5.70	*tum, mirum* dictu, commotum serpere lignum
5.679	praesentis vitae *spatium dum* ceditur aevo
6.72	spemque metumque inter, *quamquam iam* libera voti

$$\varepsilon A$$

| 6.251 | *pro quo* respondens confestim gratia praestat |

$$\zeta A$$

1.37	quaeque negant *nobis, illis* dant umida sedem
3.176	limo *formatus rursus* redigeris in arvum
3.259	ut docet *eventus) sinibus* conspexit ovantem
4.9	per quos *immissus rebus* vix paene creatis
4.553	*interea magna* pontus se mole movendo
6.93	eximiumque *decus, cuius* tu iure propinqua
6.604	in tormenta *rapi, veri* quam nescius ante

$$\zeta B$$

| 2.43 | nomen et *aeternam ponam* super aethera sedem |

AUSONIUS

*Cupido cruciatus, Eclogae, Ephemeris, Epigrammata,
Epitaphia, Epistulae, Mosella, Ordo nobilium urbium,
Parentalia, Praefationes, Professores* (3267 lines)

	A		B	
α^1	2	0.06	2	0.06
α^4	1	0.03	1	0.03
α^7	1	0.03	–	
β^1	1	0.03	–	
γ^1	–		2	0.06
γ^2	3	0.09	–	
δ^1	16	0.49	3	0.09
ϵ	3	0.09	–	
ζ	7	0.21	8	0.25

His counts on the whole are fairly close to Claudian's, low for a post-classic
in both α and ω. His most finished composition, the *Mosella* (483 lines), has
only two homs, both δ^1s.

α^1 A

Ecl. 25.10	*ecfetos ramos* denudat flamma Nepai
Prof. 20.14	et *placidae vitae* congrua vita fuit

α^1 B

Epigr. 6.4	da *rectum casum:* iam solicismus eris
Urb. 93	hanc *ambustorum fratrum* pietate celebrem

α^4 A

Epist. 11.19	quot telios *primus numerus* solusque probatur

α^4 B

Epigr. 4.1	languentem *Gaium moriturum* dixerat olim

α^7 A

Urb. 155	huius fontis *aquas peregrinas* ferre per urbes

β^1 A

Praef. 4.22	inque *meis culpis* da tibi tu veniam

γ¹ B

Prof. 22.9	quod ius *pontificum, veterum* quae scita Quiritum
Urb. 167	diligo *Burdigalam, Romam* colo; civis in hac sum

γ² A

Epist. 19.9	et *nonas decimas* ab se nox longa Kalendas
Epitaph. 35.4	*nupsit concepit peperit,* iam mater obivit
Parent. 3.18	*doctus facundus,* tu celer atque memor

δ¹ A

Ecl. 3.2	*his demptis* nil est, hominum quod sermo volutet
Ecl. 10.24	finit, ut *a bruma* mox novus annus eat
Ecl. 23.7	*matronae quae* sacra colant pro laude virorum
Ephem. 3.36	vitemus *laqueos, quos* letifer implicat anguis
Epigr. 54.4	*has geminas* artes una Sabina colit
Epigr. 80.5	tractavitque manum *victuri, ni* tetigisset
Epist. 2.6	vel cisio *triiugi, si* placet, insilias
Epist. 3.35	sunt et *Aremorici qui* laudent ostrea ponti
Epist. 8.20	non habet emptorem, sit tibi *pro pretio*
Epist. 25.12	occurre *ingenio, quo* saepe occulta teguntur
Mos. 121	stagnorum, *querulis vis* infestissima ranis
Mos. 314	iussus ob *incesti qui* quondam foedus amoris
Parent. 8.4	moribus *ornasti qui* veteres proavos
Parent. 12.12	inque domo ac *tecto, quo* pater oppetiit
Parent. 30.6	quaeque sine *exemplo pro* nece functa viri
Urb. 16	vellet *Alexandri, si* quarta colonia poni

δ¹ B

Ecl. 15.2	conficit, a tropico in *tropicum dum* permeat astrum
Epigr. 6.2	nomen qui *proprium cum* vitio loquitur
Epist. 25.13	Thraeicii *quondam quam* saeva licentia regis

ε A

Prof. 1.12	teque canam *de te,* non ab honore meo
Prof. 12.4	aetas nil *de te* posterior celebrat.
Prof. 19.10	arbitrium *de te* sumit origo suum

ζ A

Ecl. 6.2	corpora, *sublimi caeli* circumdata gyro
Epigr. 86.2	*festinas glossas* non natis tradere natis
Epist. 12.1	*Ausonius, cuius* ferulam nunc sceptra verentur

Epist. 12.31	subsidisque *fero? moneo* tamen, usque recuses
Epist. 28.5	devotus *teneas, perstas* in lege tacendi
Parent. 9.20	*dissimilis fueris;* seu bona quod similis
Praef. 1.18	cultior, et nomen *grammatici merui*

ζ B

Ephem. 3.61	sim mihi; non *faciam cuiquam* quae tempore eodem
Epigr. 46.3	Alcida *quondam fueram* doctore secundus
Epist. 18.2	*historiam: quamquam* titulo non digna sereno
Epist. 23.5	fabula non *umquam, numquam* querimonia movit[5]
Epist. 23.51	grande aliquod *verbum nimirum* diximus, ut se
Epitaph. 20.1	*Euphemum Ciconum* ductorem Troia tellus
Prof. 1.18	non *etiam luteam* volveret illuviem
Prof. 24.3	stemmate *nobilium deductum* nomen avorum

AVIANUS

	A		B	
		Fabulae (654 lines)		
α^1	1	0.15	—	
δ^1	2	0.31	2	0.31
ζ	1	0.15	1	0.15

Very low counts. Crusius (*RE* II.2376.42) says of his elegiacs: 'die metrisch-prosodische Technik ist zwar ziemlich correct.'

α^1 A

30.10	in *varias epulas* plurima frusta secans

δ^1 A

17.10	*a trepida* fertur vulpe retenta diu
26.10	tu tamen *his dictis* non facis esse fidem

δ^1 B

33.12	*nam poenam* meritis rettulit inde suis
34.16	*nam vitam* pariter continuare solent

---

[5] Bracketed as spurious by Prete.

$$\zeta\,\mathrm{A}$$

26.11　　　　nam *quamvis rectis* constet sententia verbis

$$\zeta\,\mathrm{B}$$

24.8　　　　te fieri? *exstinctam nam* docet esse feram

AVIENIUS

	A		B			A		B	
	Aratea (1878 lines)					*Descriptio orbis terrae* (1393 lines)			
α^1	3	0.16	–		α^1	2	0.14	–	
α^2	2	0.10	–		α^7	1	0.07	–	
α^5	1	0.05	–		γ^1	1	0.07	1	0.07
α^7	2	0.10	–		γ^2	1	0.07	–	
β^1	1	0.05	–		δ^1	10	0.72	3	0.22
γ^1	2	0.10	–		ζ	4	0.29	2	0.14
γ^2	1	0.05	–						
γ^3	–	–	1	0.05					
δ^1	19	1.01	3	0.15					
ζ	5	0.27	–						

Aratea

$$\alpha^1\,\mathrm{A}$$

335　　　　haec ubi *permaesto rauco* congesserat ore
630　　　　*praestantis iuvenis*, pecudes qui et flumina vates
1040　　　defessos *longo spatio* tener amputet aer

$$\alpha^2\,\mathrm{A}$$

911　　　　nullus eas *aliquo pacto* deprendere certet
1219　　　amborum capita et *palmas geminas* Ophiuchi

$$\alpha^5\,\mathrm{A}$$

385　　　　hos dixere *Asinos ortos* Thesprotide terra

$$\alpha^7\,\mathrm{A}$$

830　　　　sunt *mediae, flammae* steriles[6] ac lucis egenae
1233　　　hauserunt *pelago, toto* lepus occidit astro

---

[6] So I make it; not *mediae flammae, sterilis.*

β¹ A

1836 certa *suis studiis* affixit signa futuri

γ¹ A

1253 totaque tum *pelago caelo* descendit ab alto
1265 vasta simul *recipit persistit* pectore celso

γ² A

1050 tum duo sunt *pisces aries* taurus geminique

γ³ B

1348 *interiectorum numerum* quoque nosse dierum

δ¹ A

448 rursum *declivi si* visum tramite vergas
508 tum celer ille Aries, *longi qui* limitis orbe
594 monstrari effigiem. *diros hos* fama cometas
722 transierit, celso *late se* cardine pandit
804 tenditur *effusi vi* gurgitis. huc quoque cristae
805 cedit apex, *summa qua* lux pistrice coruscat
872 *his signis* austri raptabunt flabra fluentum
937 sidera *nocturni si* suspectare libebit
1208 ac memor *has poenas* dolor exigit. omnia fluctus
1333 ast orbis *medii si* caedat Cynthia formam
1388 vitabunt *alii, si* certis singula signis
1407 quos det certa dies, *qui longi* tempore mensis
1460 cornibus ingreditur, *si quarti* sideris ortu
1574 ipsa dei cedunt *blandi si* lumina sollers
1597 omnibus *his signis* in terram defluit imber
1599 visa deum. haec *celeri si* praesurrexerit ortu
1602 at *matutini si* Phoebum litoris acta
1668 at regione *noti si* lucem stella senescat
1694 si regione *noti, si* lenis parte favoni

δ¹ B

173 protinus, expertem *quam quondam* dixit Aratus
1741 tempore, *tum proprium* modulatur noctua carmen
1742 *tum vespertinum* cornix longaeva resultat

ζ A

431   *oceani; proprio* Taurum deprenderis ore
443   donavitque *polo. tergo* Cynosuridos ursae
955   atque sinistro *umero cubito* tenus hic quoque dextram
1337   effluxisse *sibi medii* iam tempora mensis
1501   durabunt *caelo. medio* quae edixerit ore

Descriptio orbis terrae

α¹ A

129   usque in *saxosi Pachyni* iuga. plurimus inde
311   *Romanas aquilas* Rhodanus tremit, Italidum vi

α⁷ A

1338   axe *noti Rubri* late salis obiacet unda

γ¹ A

1107   turgescit *vento, zephyro* sinus aestuat alter

γ¹ B

812   pars *Asiam, Libyam* pars adiacet altera ponto

γ² A

330   corpora *proceri, nigri* cute, viscera sicci

δ¹ A

62   vix hebes *has oras* ardor Titanius afflat
367   hic urbs est Thebae, *Thebae quae* moenibus altis
389   gentes *innumerae, quae sparsae* litore longo
653   *Ionii si* quis rate rursum caerula currat
758   longa dehinc *celeri si* quis rate marmora currit
852   accipe *qui populi* circumdent denique Taurum
964   Rhebas, *coeanei qui* dissicit aequora ponti
1184   aut *qui cornigeri* ductor gregis arva pererrat
1189   brachia *Nysaei qui* palmitis ordine iusto
1212   horum *qui gelidi* succedunt plaustra Bootis

δ¹ B

423   vitam agitant. *tum caeruleum* Padus evomit antro
484   *Ausoniam. nam* qua boreali vertice ad aethram
959   surgit Halys. *tum Paphlagonum* sata longa patescunt

ζ A

1049	totam Asiam *celsi praecingi* vertice Tauri
1067	servavere *polis. populis* Phoenicibus ergo
1110	tangitur *oceani; felici* terra recumbit
1368	*succedunt; solvunt* properantes lintea Bacchae

ζ B

| 10 | carminis *auspicium, primum* memorate, Camenae |
| 533 | in iubar *eoum rursum* se pervia flectunt |

CLAUDIAN

1–10, 15–28, De raptu (8333 lines)

	A		B	
α^1	8	0.10	5	0.06
α^3	4	0.05	–	
α^6	1	0.01	–	
β^1	1	0.01	–	
β^2	1	0.01	–	
γ^1	4	0.05	–	
γ^2	–		1	0.01
γ^3	–		4	0.05
δ^1	49	0.59	17	0.20
ε	4	0.05	–	
ζ	22	0.26	8	0.10

Results, as already remarked, resemble Ausonius's. The three books *De raptu*, totalling 1172 lines, have no alphas, and Book II (424 lines) has no homs; neither has the *Epithalamium* (9–10, 363 lines), if we admit the variant *iam superas ipsam* for *ipsam iam superas* in 271. In *Rapt.* 3.312 *quid tantum dignum fleri dignumque taceri?* Heinsius' *quid tandem* deserves to be accepted. So perhaps does Gesner's *occultos odii* for *occultis odiis* in *Rufin.* 1.257 and Postgate's *informis* for *infernis* in *Eutrop.* 2.231.

α¹ A

1.106	claustraque *congestis scopulis* durissima tendunt
1.147	his ego nec *Decios pulchros* fortesve Metellos
1.178	iam parat *auratas trabeas* cinctusque micantes

7.154	per *consanguineos thalamos* noctemque beatam
15.172	artificem: *varios sucos* spumasque requirit
15.295	*Alpinis odiis,* alternaque iurgia victi
18.138	est ubi *despectus nimius* iuvat. undique pulso
28.165	palla tegit *latos umeros,* curruque paterno

$$\alpha^1\,B$$

5.234	ducitur et *siccum gladium* vagina recusat
5.464	*rimosam patriam* dilectaque pumicis antra
20.399	*Tarbigelum tumidum* desertoresque Gruthungos
22.345	*susceptum puerum* redimitae tempora Nymphae
26.306	*obscaenam latebram* pietas ignava requiret

$$\alpha^3\,A$$

3.257	*occultis odiis*[7] terror tacitique sepultos
22.11	*discussis tenebris* in lucem saecula fudit
28.129	luce, tot *amissis sociis* atque omnibus una
28.627	in latus *allisis clipeis* aut rursus in altum

$$\alpha^6\,A$$

8.171	*hortati superi,* nullis praesentior aether

$$\beta^1\,A$$

3.274	lapsuroque *tuos umeros* obieceris orbi

$$\beta^2\,A$$

28.439	subiectum *nostris oculis* et cuius agendi

$$\gamma^1\,A$$

8.79	providus; hic *fusis, collectis* viribus ille
15.324	ille *luat; redeat* iam tutior Africa fratri
20.362	quis voci *digitos, oculos* quis moribus aptet
22.45	utque hostes *armis, meritis* sic vincit amicos

$$\gamma^2\,B$$

15.12	*congressum, profugum, captum* vox nuntiat una

---

[7] See above.

γ³ B

8.214	si tibi *Parthorum solium* Fortuna dedisset
20.68	figere. *praesidium legum* genitorque vocatur
20.590	*rectorum numerum* terris pereuntibus augent
27.11	me quoque *Musarum studium* sub nocte silenti

δ¹ A

1.29	quem prius aggrediar? *veteris quis* facta Probini
1.137	quam tua *pro Latio* victricia castra laborent
3.7	consilio firmata *dei, qui* lege moveri
3.174	incluta Thessalicis, *celsa qua* Bosporus urbe
5.22	haec fatus, ventis *veluti si* frena resolvat
5.59	aspicit. *hi vigili* muros statione tueri
5.186	si tunc *his animis*[8] acies collata fuisset
5.220	*his dictis* omnes una fremuere manipli
5.269	tempestas subeunda *mihi, qui* forte nefandas
7.155	per *taedas, quas* ipsa tuo regina levarit
8.109	non dabitis murum *sceleri. qui* vindicet, ibit
8.156	donaturque *tibi, qui* te produxerat, annus
8.309	*Romani, qui* cuncta diu rexere, regendi
8.332	effossi per operta *soli. si* longa moretur
8.480	ereptas quaesivit *aquas, quas* hostibus ante
8.502	alligat; ipsa *suo pro* pignore castra laborant
15.306	experior. *volui si* quod, dum vita maneret
15.327	ingreditur, *Tyrio quo* fusus Honorius ostro
15.468	armiger *a liquida* cunctis spectantibus aethra
18.142	vertere? *qui servi* non est admissus in usum
20.39	unam *pro mundo* Furiis concedimus urbem
20.116	dum pereunt, meminere *mali, si* corda parumper
20.251	*pro Rheno* poturus Halyn. dat cuncta vestustas
20.340	mirarique *suas, quas* Bosporos alluit, aedes
20.402	*his dictis* iterum sedit; fit plausus et ingens
20.602	unaque *pro gemino* desudet cardine virtus
21.76	tu legeris tantosque *viros quos* obtulit orbis
24.347	*qui secti* ferro in tabulas auroque micantes
26.115̣	non *qui praecipiti* traheret simul omnia casu
26.161	deficimus *queruli, si* bos abductus aratro

---

[8] V.l. *his si tunc.*

26.186	litora Leucates; *ipsae, quae* durius olim
26.258	*scrutari si* vera velis, fera nuntia Martis
26.282	esse *mihi, si* fraude nova vel calle reperto
26.314	*his doctis* pavidi firmavit inertia vulgi
26.484	*pro baculo* contis non exarmata senatus
26.493	crede *seni, qui* te tenero vice patris ab aevo
26.500	*his claustris* evade, precor, dumque agmina longe
26.528	non ita *di Getici* faxint Manesque parentum
26.583	*pro Latio* docuit †gentis praeclarus Alanae
28.171	et fluvium *nati, qui* vulnera lavit anheli
28.303	evaluit transferre *Pado. pro* foedera saevo
28.392	*his annis,* qui lustra mihi bis dena recensent
28.517	non procul amnis abest, *urbi qui* nominis auctor
28.598	Romanae tutela *togae, quae* divite pinna
Rapt. 1.182	membra regens, *volucri qui* pervia nubila tractu
Rapt. 1.262	vitales utrimque *duas, quas* mitis oberrat
Rapt. 3.171	conspicit Electram, *natae quae* sedula nutrix
Rapt. 3.276	*hos animos* bonus ille sopor castumque cubile
Rapt. 3.420	*his lacrimis?* ego te, fateor, crudelis ademi

δ^1 B

1.140	sed, precor, hoc *donum cum* libertate recenti
1.156	illis, *quam propriam* ducunt ab origine, sortem
3.139	peiorem mirata *virum: tum* talia fatur
7.14	et *regnum cum* luce dedit. cognata potestas
8.165	saepe tuas *etiam iam* tum gaudente marito
8.386	in iuvenem patiensque *meum cum* fratre tuere
18.35	nudatus quotiens, *medicum dum* consulit emptor
19.36	Hesperius *numquam, iam* nec Eous eris
20.43	semina. *tum lapidum* fletus armentaque vulgo
20.514	*tum demum* patrem implorant et nomen inani
22.381	calcatam flevere *togam: iam* prata choreis
22.468	ingreditur vallemque *suam, quam* flammeus ambit
26.54	urbs aequaeva polo, *tum demum* ferrea sumet
26.316	ausaque *tum primum* tenebris emergere pulsis
28.86	Assyrias; habeat *Pharium cum* Tigride Nilum
28.152	talia *dum secum* movet anxius, advolat una
28.305	tum mihi, *tum letum* pepigi. violentior armis

ε A

19.67	sed vereor teneant *ne te* Tritones in alto
20.91	hanc amat, hanc summa *de re* vel pace vel armis

22.283	respuis oblatum, *pro quo* labente resistis
24.119	*de se* iudicium non indignatur haberi

ζ A

8.605	cui *furerent; irent* blandae sub vincula tigres
15.261	*seditio? taceo* laesi quod transfuga fratris
15.289	molitur *Stilicho? quando* non ille iubenti
15.410	hunc quoque nunc *Gildo, tanto* quem funere mersit
15.422	dictaque ab *Augusto legio* nomenque probantes
17.334	*eloquii, duplici* vita subnixus in aevum
18.71	*crudelis? generis* pro sors durissima nostri
18.311	quam necdum *meruit, scandit* sublime tribunal
18.458	vivere *Caesareo, parvo* procede sepulchro
20.156	sic eat: in *nostro quando* iam milite robur
22.224	*elusae dominae* pergunt ad limina Romae
26.20	plurima sed *quamvis variis* miracula monstris
26.366	ac *veluti famuli,* mendax quos mortis erilis
26.399	temptandas *mediis quamvis* in luctibus iras
28.137	*remigiis, scissis* velorum debilis alis
28.191	crede *mihi, simili* bacchatur crimine quisquis
28.210	oblatum *Stilicho violato* foedere Martem
28.312	desinit esse *meus. melius* mucrone perirent
Rapt. 1.66	signa? quid *incestis aperis* Titanibus auras
Rapt. 3.101	dulce *tibi, tali* quae nunc, ut cernis, hiatu
Rapt. 3.307	forte *tibi.*[9] *nosti* quid sit Lucina, quis horror
Rapt. 3.283	utraque digna *coli! tanti* quae causa furoris

ζ B

5.424	eversis *agedum Rufinum* divide terris
7.175	o decus *aetherium, terrarum* gloria quondam
17.124	illa per *occultum Ligurum* se moenibus infert
17.284	a quibus haud *umquam palmam* rapturus Arion
18.49	aucturus *pretium; fecundum* corporis ignem
20.320	munera. *militiam nullam* nec prima superbus
20.581	proditioque *ducum, quorum* per crimina miles
22.451	*quorum praecipuum* pretioso corpore Titan

---

[9] V.l. *tibi est.*

CORIPPUS

	A	B			A	B	
	Iohannis (4664 lines)				*Laudes Iustini* (1680 lines)		
α^1	11	0.24	–	α^1	24	1.44	3 0.16
α^4	–		2 0.04	α^2	1	0.06	–
α^5	2	0.04	–	α^3	2	0.12	–
α^7	8	0.17	–	α^4	–		1 0.16
β^2	1	0.02	–	α^5	2	0.12	–
γ^1	4	0.09	1 0.02	α^7	4	0.24	–
γ^2	5	0.11	–	β^2	1	0.06	–
γ^3	2	0.04	1 0.02	γ^1	6	0.36	–
δ^1	26	0.56	13 0.27	γ^2	7	0.42	1 0.06
ζ	24	0.51	10 0.21	γ^3	1	0.06	–
				δ^1	12	0.71	2 0.12
				δ^2	1	0.06	–
				ζ	13	0.77	6 0.36

Counts are notably lower in the *Iohannis* (though not low) than in the *Laudes Iustini*, composed some eighteen years later.

Iohannis

α^1 A

1.209	haud secus *Hadriacis undis,* ventisque secundis
4.307	*alipedis celeris*[10] dextra contorsit habenas
4.561	et *summae galeae* cristis conisque micantes
5.193	sic *victos populos* praefectus voce refecit
5.280	armigeris, *victos socios* virtute levavit
5.367	*contiguus rivus*[11] pedibus calcatur et amnis
6.254	consulite et *dubios animos* firmate docentes
6.676	hic *fusos socios* et iam sua vulnera cernens
6.709	*oppressos socios.* victor iam sternitur hostis
7.518	per *Libycos populos* et pacem reddere mundo
8.256	nulla pavet *solitis sacris.* tunc arma movere

[10] *alipedes celeres* T.
[11] *contiguos uiros* T: -*uo rivo* Mazzucchelli.

$(\alpha^1 \, B)$

7.35 ille putat *genitum proprium*[12) de sanguine natum

$\alpha^4 \, B$

1.88 *Parthorum dominum victum* versumque fugavit
3.29 laeta fuit. *plenam Libyam* cultamque reliqui

$\alpha^5 \, A$

3.5 circumquaque *duces vallantes* agmine denso
8.271 dum *consanguineus, genitus* de matre Latina

$\alpha^7 \, A$

2.176 iam sonipes *campis laxis* currebat habenis
2.238 *innumeros, campos* acies implesse nefandas
3.51 auctoremque *mali prisci* seu temporis iram
4.543 hic melior *pilo, curvo* nec segnior arcu
5.233 montibus in *mediis, stimulis* et verbere caudae
5.478 cum *sociis. cunctis* stridunt de partibus enses
6.41 et munite *locos. celsos* indagine montes
7.169 luctibus in *tantis animis* dolor excidit amens

β^2

2.360 *his monitis nostris* aures contunde superbas

$\gamma^1 \, A$

5.379 missile *contorto, nudo* modo percutit ense
5.432 sarcina lapsa *ruit. cecidit* lectusque lapisque
8.437 dum *veniet, laudet* fortes rebusque fideles
8.548 rumpere non *potuit, sonuit* tamen. inde furentem

$\gamma^1 \, B$

2.469 dum putat *adversum, socium* ferit improba pectus

$\gamma^2 \, A$

2.383 *nos Lazos Unnos* Francosque Getasque domamus
4.592 castus amor *pietas bonitas* sapientia virtus
5.467 atque truces *currunt, feriunt* discrimine nullo
6.79 loricas *conos clipeos* gladiosque minaces
6.435 et pietate *probant. laudant* viresque fidemque

[12) *proprio* Mazzucchelli.

γ^3 A

1.362	*vi pelagi;* duros tabulae iacuere per agros
4.439	ergo agite, et *belli socii* civesque fideles

γ^3 B

1.88	*Parthorum dominum victum* versumque fugavit

δ^1 A

1.305	*pro Petro* nunc parce *meo.' quo* nomine dicto
1.372	felix ille locus, *statio quo* tuta Latinis
1.458	Arcitenens *crebris quis* ferret fata sagittis
2.56	*qui Gurubi* montana colunt vallesque malignas
2.62	Silvaizan Macaresque *vagi, qui* montibus altis
2.65	et *qui⟨s⟩ flumineis* fontes interserit undis
2.121	conveniunt *populi, qui* currunt arte per aequor
2.360	*his monitis nostris* aures contunde superbas
3.344	coeperat. *his nostris*[13] veniens fervebat in oris
4.217	*has poenas* rebusque fidem dominisque reservent
4.278	⟨*pro*⟩ *populo* iam ferre potest. nam mentibus omnes
4.428	exsolvit meritas *pro fracto* foedere poenas
4.608	dis torquens errore *vias, quas* callidus arte
6.78	terribilesque *viros quos* fecerat aspera caedes
6.621	si morimur, *socii, si* fors suprema Latinos
6.690	*qui lateri* iungunt, iaculis volitantibus acres
6.714	tigris anhela suis, *raptos quos* forte cubili
7.126	augeturque magis. *tumidos quos* cernitis hostes
7.231	quo pater ille *modo; quo* nunc Ricinarius una
7.496	*detinui, qui* cuncta tibi secreta resolvant
7.519	Ammonis *his dictis* gentes Bellona coegit
7.525	*quos populos* nunc rere magis, non terror acerbus
7.539	ut magis *hos nostros* teneatis certius agros
8.189	his tunicas ferrum, *his, nudis* de more lacertis
8.223	ponemus mensas. *ne longe* pascite campis
8.274	*his victis* Nasamon: cuncti tua castra sequentur

δ^1 B

1.579	dixerat haec ductor, *latum cum* vocibus agmen
2.115	et mucrone potens, *saevum*[14] *dum* tendit in hostem

---

[13] See Diggle-Goodyear.
[14] *saevus* Shackleton Bailey: *saevo* Goodyear.

3.285	dum premitis gentes, *totum dum* vincitis orbem
3.292	ante magis cecidit *quam praedam* tolleret hostis
4.215	perfidiae *meritum cum* durae crimine mortis
4.463	murmure mota novo, *medium dum* tangit Olympum
6.164	*fatidicum dum* quaerit iter. vox improba tandem
6.214	pensasset *pronum cum* magno pondere corpus
6.629	dum morimur. revocate *fugam: iam* stringite ferrum
7.124	amisisse putet. *praedam quam* forte doletis
7.272	venit Ifisdaias *centum cum* milibus ardens
7.319	mentibus et *pugnam iam* iam temptare parabant
7.331	perfunditque *cutem, quem* mox exiccat iniquus

ζ A

1.35	confossum *gladiis. natis* fas non fuit ullis
1.109	cernere quod *merui nostri* virtute Iohannis
1.213	litora nulla *quatit. siluit* tunc Scylla biformis
1.234	hic dux *magnanimus securus* puppe Iohannes
1.316	iussio certa *dei: venti* venere secundi
3.116	sanguine *Vandalico video* decurrere rivos
3.163	corruit ante *pedes aries* prostratus anheli
4.372	haecine vestra *fides? tales* referuntur amici
4.569	arma movens *populi. veluti* vigilanter oberrant
4.610	Aegides *monitus; pectus* tunc ipse biforme
5.26	omina prima *suis. celsis* tunc cornibus ille
5.373	omnis ad *obsessas aetas* concurrere fossas
5.433	quo Cererem *frangit. dirupit* vincula pondus
6.101	obtulit et *munus, summus* quod more sacerdos
6.120	occubuere *viri. veluti* cadus hauriat undas
6.160	ut *vero toto* percepit pectore numen
6.223	iam *placidas, victas* iterum bellasse catervas
6.672	*magnanimus. durus* venientis comminus ictus
7.356	impediens *populos. captivos* comprimit Afros
7.456	sternere caede *viros, vivos* sed prendere certans
7.469	ducitur hinc *victus, manibus* post terga revinctis
8.24	liber *equo, pilo* feriens vel arundinis ictu
8.415	percutit acer *equo*[15] *nigro* de corpore sanguis
8.445	*militibus medius* fervens, fortesque tribuni

---

[15] *equus* Diggle.

ζ B

1.79	sed *postquam primam* vigilans ex hostibus urbem
1.166	et latet *angustum centum* sub puppibus aequor
3.263	*annorum fessum* numero casuque paventem
3.359	o male corda *novam nusquam* plangentia mortem
4.429	quid *Stutiam referam* profugum, tot partibus orbis
5.123	armatamque *manum. ferrum* tenet illa retractans
6.89	iam servire *volunt. didicerunt* corda dolores
6.664	*Marmaridum. densum* circumvolat undique ferrum
8.73	fumum *atrum primum* parvamque exire favillam
8.147	dicite. *cognoscam, quaenam* suprema voluntas

Laudes Iustini

α¹ A

Praef. 21	mater *consilii placidi* vigilantia vestris
Praef. 29	subque *pio domino* cuncti bene vivere quaerunt
1.85	*tanti consilii*, licet haec deus omnia fecit
1.193	*divinis animis* inerat dolor ille parentis
1.194	ante *pios oculos* mitis versatur imago
1.265	*divinis animis* nulli evitabilis auxit
1.327	*maturas uvas, maturas* signat olivas
2.111	*Augustis solis* hoc cultu competit uti
2.118	*Caesareos umeros* ardenti murice texit
2.154	nec *vacuis verbis* nec inanibus ista figuris
2.224	*magnis officiis* summisque laboribus aptam
2.279	cerneret ut *laetos populos* plebemque moneret
2.296	*intentos oculos* ad sedem vulgus erilem
2.311	*Augustae Sophiae* votis quam pluribus orant
2.412	per *medios populos.* postquam venere verendam
2.415	cerne *pias lacrimas*, miserorum vincula solve
3.140	tot *latos populos, duros* domitare tyrannos
3.171	extantes *latis umeris* durisque lacertis
3.229	*auratis radiis* argentea sidera vincit
3.241	aurea et *auratos conos* cristasque rubentes
3.273	*innumeros populos* et fortia regna subegit
4.247	*angelicis oculis* exaequans sidera caeli
4.292	*internis oculis* illic pia cernitur esse
4.331	*Romani populi* patres sine semine facti

α^1 B

Anast. 1	*immensam silvam* laudum, vir iuste, tuarum
1.88	*aeternam formam* laudemque et nomen habebis
2.143	*sceptrum continuum*, tempus iuge, iuncta potestas

α^2 A

| 3.158 | acciti *proceres omnes*, scola quaeque palati est |

α^3 A

| 2.71 | *oblatis ceris* altam remeavit in aulam |
| 2.430 | *dimissis populis* arcem remeavit in altam |

α^4 B

| 4.325 | imperat, hunc *ipsum solum* spe certus adorat |

α^5 A

| 1.64 | ecce tuae *proceres pulsantes* limina portae |
| 2.135 | felix *Armatus, primus* qui verba loquentis |

α^7 A

1.327	*maturas uvas, maturas* signat olivas
3.140	tot *latos populos, duros* domitare tyrannos
3.298	miscuimus *pugnas, munitas* cepimus urbes
3.393	obstamus *dominis, profugis* damus ostia servis

β^2 A

| 1.131 | *vestris consiliis* vestrisque laboribus auctum |

γ^1 A

2.287	ambo *patricii, dilecti* principis ambo
2.318	nunc simul *erectis, pronis* nunc ardua membris
3.69	*incipiunt: plaudunt* pedibus mollique reponunt
3.169	ense latus *cincti, praestricti* crura cothurnis
3.353	sponte damus *dignis, indignis* sponte negamus
4.20	rarescunt *luci, campi* spoliantur opacis

γ^2 A

1.169	*continuit fovit monuit* nutrivit amavit
1.252	*cognatos, famulos* et tantos linquis alumnos
1.260	dispositi, *classes acies* exercitus arma
1:309	urguet agit *stimulat pulsat* latus imprimit instat

3.74	*saltatus risus discursus* gaudia plausus
3.226	*fulmineus cautus* vigilans noctesque diesque
4.282	*coepit perfecit* donisque ornavit et auxit

$$\gamma^2 B$$

Anast. 26	summe *magistrorum, procerum* decus, arbiter orbis

$$\gamma^3 A$$

4.146	consulis et *mundi domini* donisque superbi

$$\delta^1 A$$

Praef. 28	principe *pro iusto* Romanum nomen amatur
Praef. 47	huic ego *sananti si* qua est fiducia servis
Anast. 44	*hi sacri* monstrant apices. lege, summe magister
2.28	quas tibi persolvit *tanto pro* munere grates
2.125	quasque *a Vandalica* Belisarius attulit aula
2.382	reddenda est, vivus *patri qui* substitit heres
2.416	matribus his natos, *his nuptis* redde maritos
3.51	contrahitur *ripas, gelidas quas* linquere terras
3.306	intemerata *tibi, si* mavis pacta manere
4.160	qui vitam, *tanti qui* principis acta reponunt
4.168	fontis *Niliaci si* quisquam aut hauriat undas
4.348	*quos fidos* habui, mihi qui nocuere necabit

$$\delta^1 B$$

Anast. 39	et *vestram iam* sentit opem gaudetque, quod ampla
3.84	hanc dicunt carmenque *novum cum* plausibus addunt

$$\delta^2 A$$

2.324	vertice fecundos *huc illuc* flectere ramos

$$\zeta A$$

Anast. 51	principis *invicti felici* carmine dicam
1.314	solis honore *novi grati* spectacula circi
1.322	nam *viridis veris,* campus ceu concolor herbis
1.345	huc omnes *populi, pueri* iuvenesque senesque
2.335	contulit ista *deus. nullus* sua gaudia turbet
3.51	contrahitur *ripas, gelidas quas* linquere terras
3.182	aurea *convexi veluti* rutilantia caeli
3.260	hunc Avares *alii simili* terrore secuti
3.323	tu velut *ignaris falsis* rumoribus audes
3.355	exaequare *meis? nostris* non fidimus armis

3.365	*deicimus, cuius* populos pietate tuemur
4.14	emptorum *numero. studio* ruit agmen emendi
4.181	namque illas *donis conscriptis* patribus aequos

ζB

2.65	Iustini *rerum. nostrum* caput, inclita, serva
2.339	*tranquillam faciam* securis civibus urbem
2.391	tollitur in *caelum populorum* clamor ovantum
3.25	tempus in *aeternum sacrum* servantia corpus
3.249	intrantes *plenum populorum* milia circum
4.313	egrediens *templum primum* sublime petivit

CYPRIANUS GALLUS

I *Genesis* (1498 lines)

	A		B	
α^1	1	0.07	4	0.27
α^2	1	0.07	–	
α^5	2	0.14	1	0.07
γ^2	1	0.07	–	
δ^1	14	0.93	9	0.60
ζ	6	0.40	5	0.33

α^1 A

551	*Cenaeos populos* Cenezaeosque tenebunt

α^1 B

304	quae super *aequoreum campum* defessa volatu
560	*caesarum pecudum* visa est delambere carnes
973	aetherias *tenerum Iosephum* gignit ad auras
1367	et *geminum pretium,* consertus ne foret error

α^2 A

1084	ipse *deos nullos* Terebinthi abscondit in antro

α^5 A

195	*largitus dominus,* hoc se poscente rogari
1390	et veluti *notus Beniaminus* non foret ipsi

α^5 B

811	compositos *fratrem nitentem* sumere partus

γ²A

552	*Chedmoneos, Chattos* iuncta cum gente Phereza

δ¹A

79	*atqui si* studeas mellitos carpere victus
126	*his actis* dominus trepidis dat taedia vitae
276	allapsu maiore *paro, quo* grandior undis
412	sed deus, *electi qui* sensum nosceret Abrae
480	quod potuit superesse *neci; qui* nomine vero
500	accipit optatum *toto pro* munere fratrem
524	munere *pro tanto, quo* natum gignat inerti
652	*his actis.* auferre procul sua pignora iussus
755	*his actis* vetulo decedit corpore Sarra
941	incolumesque videt *cunctos quos* mente quaerebat
946	involvens lapidem, *putei qui* texerat ora
1105	visurus *sanctas, quas* dat prudentia, sedes
1239	*his dictis* sat vera quidem, sed dura loquendo
1388	confirmat cohaerere *deo pro* munere vitae

δ¹B

214	talia disseruit, *dum vatum* more futura
357	id pietatis opus *postquam iam* mente serena
683	principis arbitrio (*germanam nam* fere vates)
817	incedens claro *dominum cum* lumine vidit
854	illic perspicuo *dominum cum* lumine cernit
882	namque senex, iuvenem *totum dum* dextera lustrat
1096	conspicuum *Danem, quem* Naptalimus adurget
1248	namque videbatur, *fluvium dum* spectat amoenum
1491	ipse *etiam postquam iam* centum triverat annos

ζA

39	coniugibusque *suis positis* cum sedibus haerent
243	ius delere *mihi mundi* peccata nocentis
636	illic pro *foribus Lodus* de more sedebat
751	sistitur, *evinctus manibus* post terga retortis
1133	quem *iuxta prona* fratrum cervice ruebant
1413	*permotus precibus* vates discedere cunctos

ζB

333	sed *croceum tantum* curvandum in nubibus arcum
674	nescius ut *vetitum natarum* nosceret usum

831	ne *sociam quisquam* temptet raptare prophetae	
1491	ipse *etiam postquam iam* centum triverat annos	
1496	et revehant *secum veterum* condenda sepulchris	

DISTICHA CATONIS (290 lines)

	A		B	
α^1	1	0.34	–	
γ^1	1	0.34	–	
δ^1	4	1.38	–	
ε	5	1.72	–	
ζ	4	1.38	1	0.34

α^1 A

2.11 b	*lis verbis minimis* interdum maxima crescit

γ^1 A

4.39 b	laedere qui ⟨*potuit*⟩, *poterit* prodesse aliquando

δ^1 A

1.4 b	conveniet *nulli, qui* secum dissidet ipse
1.33 b	*pro lucro* tibi pone diem quicumque sequetur
2.11 b	*lis verbis minimis* interdum maxima crescit
4.48 b	fac discas, multa *a vita* te scire doceri

ε A

1.11 b	sic bonus esto bonis, *ne te* mala damna sequantur
1.14 b	plus aliis *de te* quam tu tibi credere noli
1.17 b	conscius ipse sibi *de se* putat omnia dici
2.12 b	quid statuat *de te*, sine te deliberat ille
3.7 b	exemplo simili *ne te* derideat alter

ζ A

1.25 a	quod dare non *possis, verbis* promittere noli
2.6 a	quod nimium est *fugito, parvo* gaudere memento
4.27 a	discere ne *cessa: cura* sapientia crescit
4.37 a	tempora longa *tibi noli* promittere vitae

ζ B

2.11 a	*adversum notum* noli contendere verbis

Late antiquity

DRACONTIUS

Romulea (2354 lines)

	A		B	
α^1	1	0.04	1	0.04
α^3	1	0.04	–	
α^5	1	0.04	1	0.04
γ^1	6	0.25	3	0.13
γ^2	4	0.17	–	
γ^3	1	0.04	1	0.04
δ^1	14	0.60	4	0.17
ε	1	0.04	–	
ζ	5	0.21	2	0.08

Orestes (974 lines)

	A		B	
α^1	3	0.30	–	
γ^1	–		1	0.10
γ^3	–		2	0.20
δ^1	7	0.72	2	0.20
ε	2	0.20	–	
ζ	3	0.30	4	0.40

De laudibus Dei (2327 lines)

	A		B	
α^1	6	0.26	1	0.04
α^2	1	0.04	–	
α^5	1	0.04	1	0.04
α^7	1	0.04	1	0.04
β^1			1	0.04
γ^1	9	0.43	1	0.04
γ^2	9	0.43	–	
γ^3	1	0.04	1	0.04
δ^1	21	0.90	7	0.30
ε	2	0.08	1	0.04
ζ	11	0.47	8	0.35

Satisfactio (316 lines)

	A		B	
β^1	1	0.31	–	
γ^3	–		1	0.31
δ^1	1	0.31	–	
ζ	4	1.23	–	

The lower counts in the secular poems are to be noted.

The 290 hexameters of *Aegritudo Perdicae* (not on my list) have an α^1 of sorts (232 *huc etiam tenerae sanctae venere puellae*), an α^6 (226 *hoc visum placitum matri*), and two ζs (83 and 281).

Romulea

α^1 A

10.503   ipse *pias animas* mittis et claudis in aevum

α^1 B

10.61   mitte *pharetratum puerum,* mea Cypris, Amorem

α^3 A

2.68   hoc *puero viso* Nympharum turba calescat

α^5 A

2.550   ignibus Idaliis *exustas Herculeas* spes

α⁵ B

10.277   et volucrem *puerum populatum* templa Dianae

γ¹ A

5.139   Taurica *crudelis, mitis* tamen, ara Dianae
5.312   gramina non *tangunt, feriunt* sed fulmina quercus
7.29   percutiant *palmis, digitis* sub arundine ventos
8.288   conubium *regni, thalami* consortia casti
10.160   purpura quod *mollis, tenuis* quod sericus ornat
10.598   Vulcanus *Lemno, Iuno* spernatur ab Argis

γ¹ B

9.63   sanguine *Troiugenum, Graium* dotata cruore
9.191   et post *vindictam veniam* concede perempto
9.216   sed veniat *tantum quantum* pensabitur Hector

γ² A

2.24   ipse sui, *satyrus cycnus* Latonia serpens
5.35   Sarmata Persa *Gothus Alamannus* Francus Alanus
7.72   *sollicitus tabidus* temerarius anxius audax
9.113   inde *homines volucres pecudes* et cuncta necantur

γ³ A

8.615   prima *ratis iuvenis* regali praedita signo

γ³ B

10.185   *telorum strepitum* castae putat arma Dianae

δ¹ A

2.140   *his dictis* mentem pueri mulcebat amica
2.159   quod matri narrabo *tuae, quae* te mihi parvum
3.17   Romuleam laetus *sumo pro* flumine linguam
3.18   et pallens *reddo pro* frugibus ipse poema
4.41   *quamvis sis* frater. iam tu succurre, Minerva
5.29   obvius armato *iuveni si* civis adesset
5.226   hoc cupit et pauper, *pro voto* divitis instat
6.38   porrexere piam *placido pro* tegmine dextram
8.25   laudis habent meruisse *cibos quos* pasta recusant
8.208   qua procul *a villa* fumantia tecta viderem
8.597   *his dictis* gemit aula ducis sub luctibus atris
9.69   *quis hostis* Diomedis erit, quae turba ferentem

10.530   sed postquam *solos quos* iusserat ignis adussit
10.548   transegit *pueros. quos* sic portabat ad aram

δ¹ B

5.47   ut legem premat ipse *suam, quam* sanxerat ante
10.343   nec gemitus latuere *magam: 'quam,* callide, fraudem
10.508   ignea mors rapiat *sponsum cum* paelice busta

ε A

10.214   sed memor esto mei, *ne te* fortuna superbum

ζ A

4.18   *compellor. genitor,* te in nosmet pessima coniunx
5.138   sit tua *terribilis, Phalaris* ceu taurus, imago
5.322   ut Thebis *partus, magnus* cum Castore Pollux
7.118   se velut *ostendi regni* de stirpe creatum[16]
10.238   impia turpis *anus. rursus* conclamat Iason

ζ B

2.134   non est flere *tuum. mundum* tibi nullus ademit
8.450   *praeceptum dum* carpit iter festinus ad urbem
10.105   cinnama cui *folium nardum* tus balsama amomum

Orestes

α¹ A

21   *extinctos titulos* victriciaque arma sepulta
53   haeret et *attonitos oculos* in virgine figit
199   atque *Agamemnonio regno* potieris et aula

γ¹ B

895   *sacrilegum superum, perfusum* sanguine matris

γ³ B

895   *sacrilegum superum, perfusum* sanguine matris
974   atque *usum scelerum* miseris arcete Pelasgis

---

[16] *velut* (n) *ostendit* Buecheler.

δ¹ A

116 *qua visa* cessere loco matresque nurusque
311 Troianas raptaret *opes spes* una manebat
549 ore fremunt *famuli, qui* carpere dentibus optant
685 viderat Atrides *muros, quos* liquerat infans
769 accipito *inferias quas* offero. victima iusta est
772 nam iacet ille *loco, quo* tu percussus obisti
917 *dis superis* grates, quod post tot funera mentis

δ¹ B

70 *nam vivam* te membra docent actusque fatentur
347 crimine cum gentis, *Danaum cum* clade suorum

ε A

441 manibus eripuit, pia coniugis, impia *de se*
678 complicibus scelerum, *ne se* subducere possent

ζ A

439 crimen *adulterii geminasti* caede mariti
667 ito *prior, senior.* nos festinare necesse est
837 dixit et angue *rogi crepitanti* pectora turbat

ζ B

610 nec *metuam quemquam.* vestri sunt patris alumni
732 matris in *exitium famulorum* nixus Orestes
776 macto *Clytaemestram, matronam* regis Egisti
817 repperit *Aeacidem subientem* templa deorum

De laudibus Dei

α¹ A

1.712 qui scit quo *nitidus crystallus* ventre creatus
2.328 audet in *externos genitos,* quam mater agit rem
3.137 debeat *aeterno domino,* qui cuncta creavit
3.184 *illaesos pueros* expavit Persa tyrannus
3.327 atque *pudicitiae laesae* castissimus ultor
3.533 ecce *deus verus,* de quo nil fingitur, in quam

α¹ B

3.406 una manus *tantum bellum* compressit inermis

$$\alpha^2\,A$$

1.121   lux facies rerum, *dux lux cunctis elementis*

$$\alpha^5\,A$$

2.592   vel quicumque *dei ficti* sermone vetusto

$$\alpha^5\,B$$

3.238   fulmina *mentitum percussum* fulmine vero

$$\alpha^7\,A$$

3.160   omnia *perpetui*[17] *generi* manifestius Isac

$$\alpha^7\,B$$

3.287   obtruncat *socium, carum* prosternit amicum

$$\beta^1\,B$$

3.186   plenius atque *suum solium* iubet omnis adoret

$$\gamma^1\,A$$

1.390   somnus erat *partus, conceptus* semine nullo
1.596   et modo quae *gelidis, calidis* nunc flatibus ora
2.101   nam quicumque *sapit, novit,* quia sic tulit artus
2.362   immemor *auctoris, mortis* dux, terminus aevi
2.556   inde reversus *abit: repetit* sua regna triumphans
2.749   mercedem *iustis, iniustis* munera nobis
2.759   contentusque *suis alienis* non inhiarit
3.304   ignibus *aetheriis, gelidis* obsessa cerastis
3.663   grande vel *exiguum, medium* minimumque profundum

$$\gamma^1\,B$$

2.31   spes *hominum, rerum* princeps mundique superstes

$$\gamma^2\,A$$

2.2   *inventor genitor nutritor* rector amator
2.26   *oceano mundo* vel caelo teste probatur
2.71   innumera, *caelis elementis* fluctibus astris
2.105   spiritus *immensus sanctus* bonus arbiter index
2.237   te fera, te *pisces pecudes* armenta volucres

[17] *perpetue* Arevalo.

2.439   taedia bella *lues clades* iactura labores
2.616   pectoris affectu secreto, mente *fide spe*
2.766   vestibus *expensis epulis* de more Tabitha
3.361   hostis erat *generis, prolis* mors, sanguinis haustor

γ^3 B

2.768   numquid ad *annorum numerum* cuicumque modesta

δ^1 A

1.167   herba virens *prodit, it* surculus omnis in auras
1.329   omnibus *his genitis* animal rationis amicum
1.354   *quo merito* sibimet data sit possessio mundus
1.396   omnia pulcra gerens, *oculos os* colla manusque
1.497   *sacrilegos quos*[18] iura dei calcasse profanant
1.525   nec mirum, *Christi si* sentit imago futurum
2.96   de pietate *sua, qua* non vult perdere mundum
2.473   *captivos quos* membra tenent et corporis usus
2.523   *supplicio quo* digna fuit, cruce verbere ferro
2.544   haec augmenta dabant; *animas, quas* claustra tenebant
2.715   *hi soli* pereunt, semper quos esse profanos
3.8   qui *numeras cunctas quas* praefert litus harenas
3.141   temporis *exigui si* sint dispendia vitae
3.254   nec tamen aeternum *modico pro* tempore quaerat
3.261   aut *certe de* strage sua. Menoecea Creontis
3.362   quid Torquata manus? *nato pro* laude periclum
3.544   *nos genitos* vocitare *suos: nos* ergo fideles
3.619   et lacrimas intende *meas, quas* fundo diurne
3.731   non ira, nam *si domini* gravis ira fuisset
3.750   sit requies *animae, quae* mox purgata quiescat

δ^1 B

1.206   ⟨*auroram*⟩ *iam* quarta dies praemiserat undis
1.408   vel dum terra *fretum, dum* terram sublevat aer
2.583   et meruit *plenam quam* contulit ante coronam
3.422   odia: constipans *magnum cum* plebe senatum
3.581   qui negat, ipse sibi *veniam iam* sponte negavit
3.664   da mihi *iam veniam*, finem concede malorum
3.733   ne peterem *veniam, quam numquam*, sancte, negasti

---

[18] Reading doubtful.

ε A

1.121	lux facies rerum, *dux lux cunctis elementis*
1.357	et *de se* tacitus quae sint haec cuncta requirit

ε B

2.188	et speciale iubes *tam quam* generale tueri

ζ A

1.709	grandinis atque *nivis venientis* quae sit origo
1.713	candida *materies, glacies* imitatur aquarum
2.364	omnibus atque *suus, solus* sub tempore parvo
2.413	pessima vota ⟨*deus*⟩, *rursus* ne perderet orbem
2.472	quid homines *miseri, fragili* sub tegmine carnis
3.8	qui *numeras cunctas quas* praefert litus harenas
3.303	solis adusta *rotis, nigris* infecta venenis
3.329	latro *maritalis, genialis* praedo pudoris
3.348	causa *pudicitiae naturae* iura diremit
3.560	impleat ipse *dei. simili* dicione voluntas
3.563	cum nos iussa *dei fieri* contemnimus ultro

ζ B

1.455	deliciis *hominum tantum* constructus opacis
2.144	miremur *dominum tantum* potuisse polorum
2.244	ut vitas *hominum tantum* mors una tulisset
2.356	invasura *fretum, tantum* contenta moratur
2.405	impia gens *hominum scelerum* mox vota resumit
2.445	sponte parans *fructum vinum* de fontibus iret
3.310	est ibi *sertorum, quorum* sub iure tenentur
3.654	ille ⟨*ego*⟩ qui *quondam retinebam* iura togatus

Satisfactio

β¹ A

49	ipse *meo domino* deus imperat atque iubebit

γ³ B

97	*Israhelitarum populum* sic culpa tenebat

δ¹ A

143	temnit praedo *cibos, quos* non facit ipse cadaver

ζ A

81	sol dat *temperies,*[19] *species* gratissima mundi
129	*conservas animas,* victum super ipse ministras
159	*confessus facinus* veniam pro clade meretur
310	*etsi peccavi,* sum tamen ipse tuus

ENNODIUS

Carmina (1500 lines)

	A		B	
α^1	5	0.33	3	0.20
α^5	1	0.07	–	
γ^1	1	0.07	–	
γ^2	10	0.67	1	0.07
γ^3	–		1	0.07
δ^1	6	0.40	3	0.20
ε	2	0.13	–	
ζ	2	0.13	1	0.07

α^1

1.5.44	et *siccis oculis* respexi marmoris iras
1.9.29	filius *excelsus dominus* deus omnia Christus
2.54.2	*communis generis,* satius sed dicitur omnis
2.76.4	et bellum *blandis principiis* geritur
2.77.10	sic *teneras culpas* qui tacet insequitur

α^1 B

1.5.27	regnator *Ligurum fluviorum* maximus ille
2.11.2	cui *faciem veterem* lux nova composuit
2.103.4	fictaque nec *verum coitum* dant ligna iuvencae

α^5 A

2.56.5	marmora *picturas tabulas* sublime lacunar

γ^1 A

1.4.60	primaevi *tremulos, fractos* imitantur ephebi

---

[19] 'Puto *temperiem*' Vollmer.

γ²A

1.1.47	ecce *Saturninus Crispinus* Daria Maurus
1.1.48	*Eusebius Quintus* gaudia magna parant
1.4.17	cui sanguis *census genius* mens vota supersunt
1.4.94	suspiret *cupiat discurrat* ferveat oret
1.9.87	*elegit voluit meruit* suscepit amavit
2.33.3	accipis *exposcis praesumis* renuis ardes
2.56.5	marmora *picturas tabulas* sublime lacunar
2.69.9	*mendicus vetulus timidus* confusus anhelus
2.80.5	pervigil *interitus ieiunus* providus ardens
2.84.5	mitis *compositus largus* pius integer audax

γ²B

2.109.9	sublimem *validum formosum* passibus aptum

γ³B

2.87.4	*virtutum pretium*, forma pudicitiae

δ¹A

1.4.101	huc, Venerem culture, *veni: qui* tardius arsit
1.9.56	servitio laudanda *suo. pro* quantus ubique
1.9.114	haec ait et *lacrimas quas* promunt gaudia fundit
2.11.9	nubila *viperei qui* gestat corda veneni
2.24.1	di, prohibete *minas, quas* partus nuntiat horrens
2.129.2	mittitur *a domina* quidquid habent pretii

δ¹B

1.4.55	*iam nusquam* Cytherea sonat, ridetur Amorum
2.10.6	viscera *dum lapidum* fingit imaginibus
2.144.3	ne foedes *cultum, dum* reddere, Maxime, non vis

εA

1.4.15	*de te* quod vernat sortitur, Maxime, mundus
2.146.4	*de.te* quandoque gaudia certa, puer

ζA

2.98.4	*gestandus manibus* saevit leo blandior ira
2.136.3	algida qui *Tanais domuistis* dorsa profundi

ζB

1.6.30	tunc *patriam teneam,* tunc stationis opem

EPISTULA DIDONIS (150 lines)

	A	B
Anth. Lat. 71 Bailey (83 Riese)		
ζ	–	1 0.67

ζ B

93   praecipitare *viam, poteram* crescentis Iuli

JUVENCUS (3211 lines)

	A		B	
α^1	20	0.62	21	0.65
α^2	2	0.06	1	0.03
α^3	2	0.06	–	
α^4	4	0.12	–	
α^5	2	0.06	2	0.06
α^7	4	0.12	–	
β^1	1	0.03	1	0.03
β^2	1	0.03	1	0.03
γ^1	2	0.06	–	
γ^2	–		1	0.03
γ^3	1	0.03	5	0.15
δ^1	23	0.72	15	0.45
ζ	18	0.56	4	0.12

This relatively early poet of Christianity has an α^{1-6} count much higher than any other on the same list except Symposius and the Laudes Iustini, not much lower than his omega.

α^1 A

Praef. 27	*dulcis Iordanis,* ut Christo digna loquamur
1.180	omnia *nocturnis monitis* quod vera recurrant
1.337	nunc ego *praeteritas maculas* in flumine puro
1.523	si te forte *oculi dextri* laqueaverit error
1.562	aut *caecis odiis* inimicos ducere dignos
1.578	conveniet *iustis meritis* tunc digna rependet
2.37	*infidos animos* timor irruit. inde procellis

2.513	tune *piis animis* requies, quam regia caeli
2.705	hic et *Ionaeis monitis* potiora iubentur
2.706	contemnitque *feris animis* gens impia lucem
3.145	me populus *summis labiis* sublimat honore
3.514	prendere *praecelsis meritis* fastigia vitae
3.665	sed *veris verbis* iterumque iterumque monebo
3.758	*progressi famuli* per compita cuncta viarum
3.765	isque ubi *regalis sermonis* pondere causas
4.58	*accubito*[20] *primo* cenae fastuque superbo
4.182	sed vos *intentis animis* assistite semper
4.196	illum *perpetuus fletus* stridorque manebit
4.247	*nequitiae tantae* veniam concedere possem
4.513	pars *strictis gladiis,* pars fidens pondere clavae

$$\alpha^1\,\mathrm{B}$$

1.234	quique *profetarum veterum* praedicta recensent
1.435	exhinc per *terram Galilaeam* sancta serebat
2.73	credere *cernentum populorum* turba coacta est
2.149	increpat *ignarum sponsum,* quod pulchra reservans
2.151	*his signis* digne *credentum discipulorum*[21]
2.178	*primorum procerum* Iudaei nominis unus
2.255	dispernens *veterum Samaritum* iussa ministrem
2.304	respondit. sed *tum mirantum discipulorum*
2.549	fulgentis caeli et *terrarum frugiferentum*
2.562	praeterit, et populus *sectantum discipulorum*
2.676	quaeratur *veterum scriptorum* lectio vobis
2.809	*secretum lolium* conexo fasce iubebo
3.182	tunc etiam precibus *sectantum discipulorum*
3.259	cogit *concilium sectantum discipulorum*
3.344	cur scriptis *veterum scribarum* factio certet
3.624	hinc lectis iussit *sectantum discipulorum*
3.632	insternunt *pullum placidum* praebentque sedendum
3.684	*captantum procerum* mentem; nam magna profetam
3.711	at vos *tantorum scelerum* nil paenitet umquam
3.734	*tantorum scelerum.* sed vobis tradita quondam
4.201	ornatu accinctae *taedarum flammicomantum*

---

[20] V.l. *accubitu.*
[21] Cf. 2.304, 562. 3.182, 259, 624.

$\alpha^2 A$

3.404	qui vero e *parvis istis* deceperit ullum
3.417	ex *istis parvis* genitor sic perdere quemquam

$\alpha^2 B$

3.699	post *alium natum* simili sermone iubebat

$\alpha^3 A$

1.310	ad *deponendas maculas* clamore vocabat
2.143	*completis labiis lapidum; tum* spuma per oras

$\alpha^4 A$

1.454	*felices humiles,* pauper quos spiritus ambit
1.456	his *similes mites,* quos mansuetudo coronat
3.488	exuit; atque *alios ipsos* sibi demere constat
3.591	ut *Christo medio* caeli sublimis in arce

$\alpha^5 A$

2.649	mox me *mittentis genitoris* dona patescent
4.602	quem Christo *infensus populus* dimittere vitae

$\alpha^5 B$

2.478	vel domini *similem virtutem* prendere servo
2.611	si gemina *regnum distractum* parte dehiscat

$\alpha^7 A$

1.262	avellit *ferro nullo* sub crimine culpae
1.380	vitam *credentis facilis* substantia panis
2.808	crescere cum *lolio. pleno* nam tempore messis
4.455	donec regna *patris melioris* munere vitae

$\beta^1 A$

1.715	quisque *meis monitis* auresque et facta dicabit

$\beta^1 B$

2.776	quisque *meum verbum* summas dimittit in aures

$\beta^2 A$

3.537	quid *nostris animis* superest? dic, Christe, precamur

$\beta^2 B$

1.655	*iudicium vestrum* fugiat damnatio saeva

γ^1 A

1.635	terga soli *subigunt, iaciunt* aut semina farris
3.405	si *sapiat, nectat* saxo sua colla molari

γ^2 B

1.649	ergo *cibum, potum* vestemque et inania cuncta

γ^3 A

4.558	maiestas *prolis hominis,* cum dextera sanctae

γ^3 B

2.77	*officium quorum* morbus dissolverat acer
3.259	cogit *concilium sectantum discipulorum*
3.548	*primorum meritum* postremi transgredientur
4.35	nec Deus *illorum dominum* se ponere mavult
4.638	*eventum rerum* patefecit in ordine saecli

δ^1 A

1.175	et simul *his dictis* caeli secreta revisunt
1.265	complorat, subolis *misero pro* funere matres
1.269	exstinxisse putat *cunctos, quos* unus et alter
1.375	'si te *pro certo* genuit Deus, omnibus' inquit
1.452	*hos populos* cernens, praecelsa rupe resedit
1.461	felix, *qui miseri* doluit de pectore sortem
1.503	*qui fatui* miserive cient sub nomine fratrem
1.583	*his votis* pompae fructus succedit inanis
1.723	hunc similem faciam *volucri qui* fulcit harena
2.151	*his signis* digne *credentum discipulorum*
2.184	Christus ad haec: 'iteris *iusti si* culmina quaeris
2.339	*his verbis* fructum mox perceptura salutis
2.366	*qui sponsi* laetis comitantur vota choreis
2.537	maior Iohannis *nostri qui* viribus esset
2.761	illis *pro merito* clauduntur lumina mentis
3.6	accipite, atque homines *puro pro* semine iustos
3.126	cuncti, navigio *socios quos* casus habebat
4.164	*ni soli* rerum domino, qui sidera torquet
4.380	*his dictis* contra depromit talia Christus
4.383	iam genitoris adest, *fidei si* robur habetis
4.399	*qui tanti* Mariam fuerant Marthamque secuti
4.431	ultima *qui domini* caperet mandata, iubebat
4.763	*his dictis* visisque animos perfuderat ardens

δ¹ B

Praef. 15	quod si *tam longam* meruerunt carmina famam
1.108	turba *propinquorum, tum* gaudia mira frequentes
1.256	*Aegyptum cum* matre simul transportat Ioseph
1.438	et mox crebra procul *Syriam iam* fama tenebat
2.116	ante *etiam quam* te vocitarent verba Philippi
2.143	*completis labiis lapidum; tum* spuma per oras
2.263	hunc Iacob etenim *puteum cum* prole bibebat
2.304	respondit. sed *tum mirantum discipulorum*
2.314	cunctaque *maturam iam* rura exposcere messem
2.321	talibus alloquiis *comitum dum* pectora complet
2.677	perpetuam cunctis *vitam quam* ferre putatis
3.37	*nam quondam* cernens liventi pectore daemon
3.318	*tum secum* iubet abruptum conscendere montem
3.440	praecipit *abreptum cum* coniuge progenieque
4.634	quod dare *tum licitum,* cum sanguis distraheretur

ζ A

1.58	*salve, progenie* terras iutura salubri
1.380	vitam *credentis facilis* substantia panis
1.496	*audistis veteris* iussum moderamina legis
1.510	offer grata *deo tranquillo* pectore dona
1.512	corporis: hoc *casti celeri* curetur amore
1.669	pulsantique *aditus foribus* pandetur apertis
1.761	at tibi iam, *iuvenis, mentis* virtute fidelis
2.543	si *vultis volucris* penetralia noscere saecli
2.548	nunc tibi *confiteor, genitor,* cui gloria servit
3.245	dicitis *aversi, navi* quod tollere panes
3.285	*arbitrio, caelo* pariter nodata manebunt
3.377	istius hinc *montis possetis* vellere verbo
3.630	*discipuli celeri* complent praecepta paratu
3.760	complentur *mensae mixtae* sine nomine plebis
4.290	aut peregrina *mihi tecti* vestisve parumper
4.562	atque ait: 'audistis *pugnantis* foeda profani
4.642	traditus est *trucibus iustus* scelerisque ministris
4.759	nulla istic *iaceant, fuerant* quae condita membra

ζ B

1.21	sobrius *aeternum; clausum* quem spiritus ipsis
2.85	'hoc' inquit 'verbum scribarum* dicta retractant
3.251	quattuor et *rursum cenantum* milia panes
4.606	tum iudex *iterum procerum* disquirere mentem

LAUS HERCULIS (137 lines)

	A		B	
δ^1	1	0.73	2	1.46
ζ	–		1	0.73

δ^1 A

65   *his coeptis* non ulla parat cunabula fatum

δ^1 B

21   iam grave plus *etiam quam* ventris tempora vellent
56   gaudebas *tantam iam* tum meruisse novercam

ζ B

60   efferat *aetherium quantum* volet orbis in axem

MARIUS VICTOR

Alethias (2020 lines)

	A		B	
α^1	2	0.10	1	0.05
α^2	1	0.05	–	
α^4	2	0.10	–	
α^5	1	0.05	–	
γ^1	1	0.05	–	
γ^2	2	0.10	–	
γ^3	–		3	0.15
δ^1	20	0.99	3	0.15
ε	6	0.30	–	
ζ	6	0.30	4	0.20

Notably low in α^1s.

α^1 A

1.379   alternum *curae propriae* misceret amorem
3.594   *corporeis vitiis* dominum iussisse recisis

α^1 B

2.492   tutius in *sedem stabilem* subducta locaret

$$\alpha^2\,\mathrm{A}$$

3.226   *denos centenos quos* prisca excesserat aetas

$$\alpha^4\,\mathrm{A}$$

3.207   Leucorum *factus medicus* nunc Gallica rura
3.668   *emissos famulos* pacisque iraeque ministros

$$\alpha^5\,\mathrm{A}$$

3.495   ut per te *genitus populus* meus arva beata

$$\gamma^1\,\mathrm{A}$$

3.315   *conspexit cecidit* congestaque funera passim

$$\gamma^2\,\mathrm{A}$$

1.41   *disposuit iussit monuit* munusque creatum
2.243   arma⟨ti⟩ *telis gladiis* et vindice ferro

$$\gamma^3\,\mathrm{B}$$

Precatio 10   tu *spatium rerum,* mentis quocumque recessus
1.257   inque *hominum cultum* nudo quod nomine venit
3.87   *natorum meritum* tali mercede rependit

$$\delta^1\,\mathrm{A}$$

1.299   iustior Euphrates, *diti qui* gurgite largus
1.433   motus, *qui sciri* faceret quodcumque necesse est
2.28   insinuat iam dulce *mori, ni* maior egestas
2.47   *his oculis –* ceu mentis erant arcana tuentes –
2.56   exaudi *miseros, quos* semper cernis et audis
2.119   ignis edax penitus *sumpto quo* victa labore
2.163   forsitan et *cunctos quos* fingit opinio casus
2.167   moverit in fetus *varios, quos* imbre marito
2.261   percussoque *animo, quo* se tortore profundo
2.336   horum *de serie,* quos continet alma propago
2.548   sustinuisset *eo quo* condidit omnia nutu
3.11   libat *odoratos quos* crine exedit honores
3.17   etsi parva fuit *tanto pro* nomine poenae
3.96   mille minus decies *quinos, quos* summa recusat
3.138   subdidit *arbitrio, quo* fatum induceret orbi
3.226   *denos centenos quos* prisca excesserat aetas
3.402   tu mihi solus eris, *populi qui* regna tenebis
3.434   de grege bisque novem *famulos, quos* instruit armis

| 3.715 | quae praestare velint. *generi, si* credere digni |
| 3.734 | *a patria* virtute deo miscente coacta |

$$\delta^1 B$$

Precatio 86	erexisti *hominem, quem* saevi fraude veneni
1.389	nam *dum terrarum* vitiis et labe carerent
3.85	sublatam medio. *nam postquam* libera somno

$$\varepsilon A$$

2.66	tunc mihi vindictam *de me* praestare putarem
2.286	testis poena probet, *ne se* permisceat umquam
2.287	ammoveat maestis *ne se* parentibus umquam
2.454	nil *de se* sapiunt. ergo omni semine vitae
3.477	quem iam mente cupis, *pro quo* placet iste Damascus
3.680	et, *se ne* totam domini clementia mitis

$$\zeta A$$

1.199	vos operante *deo? tanto* quis digna parenti
2.150	admonet arte *sequi. duri* tunc pondere saxi
2.241	auctor primus *eris caedis.* te publicus hostis
2.257	ut, quia *praemonitus facinus* committere dirum
3.236	ut nos *posteritas, terras* quom liquimus istas
3.401	damna *patris tuleris* sancti innumerabile vulgus

$$\zeta B$$

1.112	ire semel *iussum, revolutum* conficit annum
Precatio 91	reddis et in *caelum tecum* vehis hoste subacto
Precatio 123	per *Iesum Christum,* qui filius unice tecum
3.15	officiis *hominum, quorum* non indiget usu

MAXIMIANUS (686 lines)

	A		B	
α^1	3	0.44	–	
α^3	1	0.15	–	
α^4	–		1	0.15
β^1	1	0.15	–	
γ^1	3	0.44	–	
δ^1	4	0.58	3	0.44
ε	1	0.15	–	
ζ	3	0.44	4	0.58

$$\alpha^1 A$$

1.12   et *veros titulos* res mihi ficta dabat
4.11   nunc *niveis digitis,* nunc pulsans pectine chordas
5.15   haec erat *egregiae formae* vultusque modesti

$$\alpha^3 A$$

5.72   *effusis lacrimis* talia verba dedi

$$\alpha^4 B$$

1.199   se *solum doctum,* se iudicat esse peritum

$$\beta^1 A$$

5.131   sternitur icta *tuo votivo* vulnere virgo

$$\gamma^1 A$$

1.65   quaeque peti *poterat, fuerat* vel forte petita
1.119   iam minor *auditus, gustus* minor, ipsa caligant
3.27   fallere *sollicitos, suspensos* ponere gressus

$$\delta^1 A$$

1.133   *pro niveo* rutiloque prius nunc inficit ora
1.235   *his dictis* trunco titubantes sustinet artus
2.73   *his lacrimis* longos, quantum fas, flevimus annos
5.46   subditus *his flammis* Iuppiter ipse fuit

$$\delta^1 B$$

1.174   *ipsum cum* rebus subruat auxilium
1.265   morte mori melius *quam vitam* ducere mortis
2.21   *quam, postquam* periit quicquid fuit ante decoris

$$\varepsilon A$$

1.207   arridet *de se* ridentibus, ac sibi plaudens

$$\zeta A$$

1.216   *diminui nostri* corporis ossa putes
3.61   *prostratus pedibus* verecunda silentia rupi
6.1   claude, precor, *miseras, aetas* verbosa, querelas

$$\zeta B$$

1.198   hoc *tantum rectum,* quod sapit ipse, putat
3.23   at *postquam teneram* rupit verecundia frontem
4.23   o *quotiens demens, quotiens* sine mente putabar
5.111   haec genus *humanum, pecudum* volucrumque, ferarum

MEROBAUDES (271 lines[22])

	A		B
γ^3	1	0.37	–

γ^3

Carm. 2.12	n)ec *numerum regum* poscere vota timent

NEMESIANUS

	A	B		A	B
	Eclogues (250 lines)			*Cynegetica* (325 lines)	
β^2	1 0.40	–	α^1	1 0.31	–
δ^1	4 1.60	3 1.20	γ^2	–	1 0.31
ζ	2 0.80	2 0.80	δ^1	1 0.31	2 0.60
			ζ	–	1 0.31

On the reading in *Ecl.* 3.6 see *Phoenix* 32 (1978).321.

Eclogues

β^2 A

4.15	cur *nostros calamos,* cur pastoralia vitas

δ^1 A

1.9	*hos annos* canamque comam, vicine Timeta
1.41	tu nostros adverte *modos, quos* ipse benigno
2.26	ex *quo consueto* Donacen exspecto sub antro
3.12	iamque videns '*pueri, si* carmina poscitis' inquit

δ^1 B

2.7	*tum primum* dulci carpebant gaudia furto
3.37	*tum primum* laetas extendit pampinus uvas
3.59	*tum primum* roseo Silenus cymbia musto

[22] Including 30 *De Christo.*

ζA

1.53	*assueras, varias* patiens mulcendo querellas
2.36	esse *mihi, nosti* numquam mea mulctra vacare

ζB

1.51	concilioque *deum. plenum* tibi ponderis aequi
3.6	haec pueri, *tamquam praedam* pro carmine possent

Cynegetica

α^1A

116   corporis et *venis primaevis* sanguis abundat

γ^2B

37   fumantemque *Padum, Cycnum* plumamque senilem

δ^1A

30   sunt *qui squamosi* referant fera sibila Cadmi

δ^1B

162	nam *tum membrorum* nexus nodosque relaxant
324	venemur dum mane *novum, dum* mollia prata

ζB

74   imbellemque *fugam referam* clausasque pharetras

ORIENTIUS

Commonitorium (1036 lines)

	A		B	
α^1	3	0.29	—	
α^2	2	0.19	—	
α^3	1	0.10	—	
β^1	2	0.19	—	
γ^1	2	0.19	1	0.10
γ^2	6	0.48	2	0.19
γ^3	—		1	0.10
δ^1	8	0.77	2	0.19
δ^2	1	0.10	—	
ε	2	0.19	—	
ζ	5	0.48	2	0.19

$$\alpha^1\,A$$

1.253	ergo *piis votis* et sanctis perfice rebus
1.331	nemo, licet *mediis muris,* sub tempore belli
1.513	et miseros *poenis variis vis* ultima cogat

$$\alpha^2\,A$$

1.41	ora *homines omnes* et muta et bruta tenebunt
2.21	ac *studiis totis* et tota nitimur arte

$$\alpha^3\,A$$

2.19	*lenito titulo* parcum se dicit avarus

$$\beta^1\,A$$

1.74	nolo *tuos hircos,* nolo *tuos vitulos*

$$\gamma^1\,A$$

1.276	poena *reos, iustos* gloria suscipiat
2.242	luminibusque *illinc, hinc* venit aure dolor

$$\gamma^1\,B$$

1.39	ergo nisi *eloquium, sensum* nisi, Christe, ministres

$$\gamma^2\,A$$

1.105	quod manibus *tangis, graderis* pede, lumine cernis
1.106	aure *audis, sentis* naribus, ore probas
1.217	id *maestis dubiis trepidis* labentibus offer
2.147	adde thymum *violas casias* melilota crocumque
2.235	*incurvos querulos* consumpto corpore numquam
2.267	robore *famosi, laudati* munere formae

$$\gamma^2\,B$$

1.11	*lascivum miserum* fallax breve mobile vanum
2.377	inter et *infantum matrum* iuvenumque senumque

$$\gamma^3\,B$$

2.313	hos inter *rerum dominum* quicumque negarit

$$\delta^1\,A$$

1.191	quid mirum *domini si* iussa verenda secutum
1.513	et miseros *poenis variis vis* ultima cogat
1.542	ut doluere *illi qui* tenuere prius

2.11	*quo studio* nostri servabis verba libelli
2.26	*pro domino* parvus, magnus apud dominum
2.84	*ollis quis* nihil est, hoc, tibi quod superest
2.149	cum tot in *his terris* peccator munera sumas
2.340	*pro merito* vivunt nunc bene, post melius

$$\delta^1\,B$$

| 1.521 | nec solus; *sociam nam* foedi in crimine falsi |
| 2.274 | *quam poenam:* factis congrua poena manet |

$$\delta^2\,A$$

| 2.63 | cum caput *huc illuc* vergat gressusque vacillet |

$$\varepsilon\,A$$

| 1.28 | ac per *te de te* sit tibi sermo placens |
| 2.411 | quod *ne me* primis credas iniungere labris |

$$\zeta\,A$$

1.71	omnia sunt *eius cuius* nos esse fatemur
1.169	nec tamen haec *dominus, cuius* sunt omnia, quaerit
1.529	ut pater *alterius factus* credatur et heres
1.582	crede *mihi fieri* vel mihi non fieri
2.17	omnia dum *volumus, facimus* quaecumque, probari

$$\zeta\,B$$

| 1.359 | hunc quoque *laudatum psalmorum* carmine David |
| 2.85 | sentio *iamdudum tacitum* te dicere, lector |

PALLADIUS

	A		B
De insitione (170 lines)			
α^1	1	0.59	—
α^7	1	0.59	—

$$\alpha^1\,A$$

| 150 | *discissi pruni* cortice fixa tegit |

$$\alpha^7\,A$$

| 105 | aemula dura *piri despecti* mala saporis |

PAULINUS OF NOLA

6, 15, 16, 18–22, 27, 28, 31, 32
(5349 lines)

	A		B	
α^1	16	0.30	5	0.09
α^2	2	0.04	1	0.02
α^3	2	0.04	–	
α^4	1	0.02	–	
α^5	1	0.02	–	
α^7	1	0.02	1	0.02
β^1	4	0.07	1	0.02
β^2	2	0.04	–	
γ^1	3	0.06	1	0.02
γ^2	2	0.04	–	
γ^3	–		1	0.02
δ^1	36	0.67	22	0.40
ε	13	0.24	–	
ζ	8	0.15	5	0.09

α^1 A

16.9	*innumeri populi,* quo me specialia tangunt
16.12	in *domino Christo*[23] sum deditus; hunc etiam oris
16.118	qua vel mole putri vel *araneolis obductis*
18.453	qui prope *caecatis oculis* tua comminus adsto
19.33	*divini veri* radios, caligine taetra
19.67	laudibus *aeterni domini* ferit aethera clamor
20.1	saepe *boni domini* caris famulantur alumnis
21.21	hostibus *Augusti pueri* victoria pacem
21.296	natus ut *aeternae vitae* puer arbore ab illa
21.420	eximis et *Christo domino* mea meque reservas
21.721	praebuit ubertim *gratis operis.* locus altis
27.119	inde *bonus dominus* cunctos pietatis ut alis
27.163	iunctus adest *domini Christi* comitatus amicis
28.173	quod nova in *antiquis tectis,* antiqua novis lex
31.155	mortem hominum *Christi crucifixi* morte subactam
31.207	ut, quia *corporeis oculis* divina teguntur

---

[23] *Domino* may be regarded as equivalent to an epithet.

$$\alpha^1\,B$$

20.171	solveret. ecce *malum servum* refugamque voracem
20.199	itur ad *hospitium notum,* deponitur illic
21.641	quaeque *animam sanctam* manet in regione superna
32.72	et *Ianum geminum* veteres dixore Latini
32.241	*supplicium proprium* timor ⟨est⟩; tormenta reatus

$$\alpha^2\,A$$

| 21.441 | si *totus mundus* mihi res privata fuisset |
| 28.23 | *binis historiis* ornat pictura fidelis |

$$\alpha^2\,B$$

| 18.419 | atque *habitum totum* spumosa per oscula foedant |

$$\alpha^3\,A$$

| 6.320 | prima tibi *dandae veniae* permissa potestas |
| 19.485 | interea *ignaris nostris* nox illa diesque |

$$\alpha^4\,A$$

| 19.532 | quod sic forte *reo capto* tunc accidit, in qua |

$$\alpha^5\,A$$

| 18.194 | quid *nobis minimis* horum praestare coronae |

$$\alpha^7\,A$$

| 6.308 | Isaiae *vatis, veteris* qui maximus aevi |

$$\alpha^7\,B$$

| 21.499 | vile bonis *pretium, pretiosum* virus avaris |

$$\beta^1\,A$$

6.309	mittam, ait, ante *tuos oculos,* o nate, ministrum
18.460	redde *meos oculos.* nam quid iuvat esse reductos
20.180	sed iam parce *tuo misero,* precor, optime Felix
20.287	inque *tuo merito* magnis insignibus altam

$$\beta^1\,B$$

| 21.378 | ante *tuum solium* quasi te carpente totondi |

$$\beta^2\,A$$

| 18.364 | quae *sociis vestris* ut praedae causa remansit |
| 20.25 | namque ad *avaritiae nostrae* lacrimabile probrum |

γ¹ A

6.17	si mors illa *fuit, meruit* quae sanguine caelum
19.204	Cypris *adulteriis, furiis* regnare Lyaeus
31.181	corpore mors *cecidit, surrexit* corpore vita

γ¹ B

31.507	contemnis *caecum, leprosum* tangere vitas

γ² A

19.396	impulit *armavit caecavit* praecipitavit
21.307	pulchra *ferax vivax* ardua odora virens

γ³ B

22.160	praebebant *dominum rerum* recinentia Christum

δ¹ A

6.58	*qui nasci* iubet hunc, idem iubet esse Iohannem
6.81	ergo ad condignas *tanto pro* munere grates
6.116	ille ait: *o toto* quem solis circulus ambit
6.123	*illi qui* caelum terras mare sidera fecit
16.124	*qui domini* tutus gremio candentia tela
16.183	cum deus ipse *suo pro* confessore coronam
18.75	sed quia non idem *tumuli qui* membra piorum
18.79	laeta bonos, cruciata *malos, quos* rursus in ipsum
18.160	et *toto quo* mundus erit fulgebit in aevo
18.170	*religio quo* crimen erat minitante profano
18.293	et capis includente *deo, quo* cuncta tenentur
18.440	in pecore. ille autem, *qui tanti* muneris alto
19.125	aut si sunt *miseri, di* non sunt atque homines sunt
19.463	sustinet. in tribus *his scyphulis* inserta relucent
19.551	nescius hoc *ipso, pro quo* fugitare parabat
19.592	vinctus *eas quas* in sacra dona tetenderat audax
19.678	intactam *ferro, quo* cetera fregerat, unam
20.139	martyris, aeger ubi *sancto pro* limine fusus
20.255	quas illi referam *tanto pro* munere digne
20.387	constitit *ignoto pro* limine quadrupes hospes
21.101	laudibus et *Domini, qui* conditor oris et artis
21.421	nam *quo consilio* rebus capitique meo tunc
21.436	Christus habet; neque tantum *isto quo* sumpserit istinc
21.526	non solis tibi *nos iunctos* vis degere tectis
21.598	*pro nardo* vasclis cumulum erumpentis harenae

21.780	proflueret damnoque *pio, quo* martyris aulis
21.788	quodnam igitur *tanto pro* munere munus, Abella
22.72	iure potens homo quisque *sui, qui* deditus uni
27.313	unde mihi *hos animos?* quae me levat aura superbum
27.350	sollicito nostros *toto quo* defuit actus
27.503	tempore *oportuno pro* peccatore rogabis
27.604	sit Felix donante *deo, quo* praesule posce
31.499	*his monitis* sanctam discamus vivere Christo
31.596	vivere, *si nostri* sis memor ad dominum
32.7	quo nos *exemplo pro* magnis parva canemus
32.13	cum duce *qui mergi* infestos vidit equestres

δ¹ B

6.87	ac *dum promissum* cunctantia corda volutant
6.288	*quamquam iam* nimius longe processerit error
15.84	Christi; sed quis *tam variam* miretur ab uno
16.72	nescio *Felicem quem* quaeritis. ilicet illi
16.256	vicit *avaritiam; nam* praedia multa domusque
18.166	hospite nobilitat *Nolam, quam* gratia Christi
18.409	*tum demum* nota specie sibi bubus apertis
19.342	*discipulum cum* fratre Petri. iam quanta per istam
19.500	et prope *iam Nolam* veniens subsistit in ipso
20.34	disparis harmoniae *quondam, quam* corpus in unum
20.358	nec sinus advexit nec mens sua *tam spatiosam*
21.736	et *totam, quam* longa dies aestate moratur
21.775	ostendit tibi rem esse *suam, quam* tu eius amico
21.790	obsequium nomenque *tuum dum* praedico signans
27.534	Ruth sequitur, *sanctam quam* deserit Orpha parentem
28.75	interea *quadam primam iam* nocte quietem
28.247	saeva vorant *animam, quam* vicerit aegra voluptas
28.272	et viles *holerum cum* sentibus eruit herbas
31.538	regis ad *occursum, dum* brevis hora patet
32.128	quid loquar et *Vestam, quam* se negat ipse sacerdos
32.134	nomine de proprio *dictam, quam* tradidit ipsa
32.238	quod *peccatorem quem* paenitet antea lapsum

ε A

6.322	*de te* Christus ait: concessum est visere talem
18.47	nullus opum, famulor *de me* mea debita solvens
18.268	quo deceptus eam? quem criminer? an tibi *de te*
19.5	ipse deus *pro quo* vitam voluere pacisci
19.124	quae divina putas. *si di* sunt, nec miseri sunt

19.266	sed magis ut pateat quia nunc *hi, qui* cruciantur
19.551	nescius hoc *ipso, pro quo* fugitare parabat
20.300	apposuere sibi *de re* superante salutem
21.689	inque petra fundare domum et *de te* bibere undam
27.302	mercari propriam *de re* pereunte salutem
27.335	quam metui *ne te* mediis regionibus hostis
31.143	sed *ne me* dubiae suspenderet anxia mentis
31.527	nemo ope confidat propria aut *de se* sibi plaudat

ζA

6.90	nuntius e *caelo famulo* tam clara referret
6.133	assuetumque *sibi facili* petit aethera nisu
6.144	et quo iussa *venit; movit* materna Iohannes
21.423	*affuerit, docuit* rerum post exitus ingens
21.733	et sub fasce *gravi cophini* cervice subacta
22.110	nemo *fugit, movit* caeli simul et maris iras
32.4	haec ego *disposui leni* describere versu
32.248	et terrore *pio rutilo* nimis igne coruscat

ζB

16.153	ingrediar, mala non *metuam, quoniam* tua mecum
22.163	et quem *cognatum*[24] *iunctum* mihi foedere laeter
28.75	interea *quadam primam iam* nocte quietem
32.32	philosophos *credam quicquam* rationis habere
32.223	remque *novam dicam* nec me dixisse pigebit

PAULINUS OF PELLA

Eucharisticos (616 lines)

	A		B	
α^1	3	0.49	3	0.49
α^3	1	0.16	–	
β^1	1	0.16	–	
δ^1	3	0.49	1	0.16
ϵ	–		2	0.32
ζ	5	0.81	3	0.49

[24] Read variant *cognato?*

α¹ A

165   cedere et *ingenuis oblatis* sponte caverem
248   ilico me *indocilis fratris* discordia acerba
310   cunctis quippe *bonis propriis* patriaeque superstes

α¹ B

46   navigeram per portam, quae *portum spatiosum*
336   armata in *caedem specialem* nobilitatis
339   instantemque mihi *specialem percussorem*

α³ A

418   *quamvis profusis dominis* nimiumque remissis

β¹ A

57   conciliare *suis meritis* potuisse videntur

δ¹ A

15   te donante, deus, *lapsi qui* temporis annos
213   nec tamen *his ipsis* attentior amplificandis
350   hac tamen *hos nostros* spe sollicitante paratus

δ¹ B

171   visum autem neque *illum tum,* quia est cito functus

ε B

480   qui fuerat, linqui et *quam iam* non posse probarem
614   me, vel in hoc proprio mortali corpore *dum sum*

ζ A

181   coniugis, antiquo *potius cuius* domus esset
216   sed potius, *fateor, sectator* deliciarum
418   *quamvis profusis dominis* nimiumque remissis
438   quae peritura *cito illo* me in tempore amasse
558   seu quod *divitibus contentus* cedere natis

ζ B

277   dissona et *interdum carorum* vota meorum
466   quem semel *invectum maiorum* traditione
578   haut equidem *iustum, verum* tamen accipienti

PAULINUS PETRICORDIAE

De vita Martini (3622 lines)

	A		B	
α^1	6	0.17	3	0.08
α^2	–		1	0.03
α^5	1	0.03	–	
α^7	2	0.06	–	
β^1	1	0.03	–	
γ^1	3	0.08	–	
γ^2	5	0.14	–	
γ^3	–		1	0.03
δ^1	12	0.33	18	0.50
ε	3	0.08	–	
ζ	17	0.48	20	0.55

$\alpha^1\,$A

1.379	*elisos oculos* et siccas sanguine fauces[25]
2.21	donec *compositis verbis* lacrimisque coactis
5.34	cum *lustris geminis* maerens annisque duobus
5.530	haut *longo spatio* praefata amotus ab urbe
6.15	comperta e *tantis titulis?* quid contigit unus
6.435	spes est una *deus solus.* quaecumque creavit

$\alpha^1\,$B

2.264	*iugem temperiem* mentito vere tegebat
2.443	celsius *insanam dextram* suspendit in altum
5.704	et *collisorum lapidum* crepitare fragorem

$\alpha^2\,$B

1.255	cumque *ipsum murum* fidei validamque columnam

$\alpha^5\,$A

5.822	tali igitur *dominus confusus* clade suorum

α^7

3.284	pro levibus *signis tumidis* plus credere verbis
5.8	ut taetrae *mortis tristis* discedat imago

---

[25] See Virg. *Aen.* 8.261.

β¹A

1.32 cresce *tuis titulis:* mage laus est nam tua vinci

γ¹A

4.375 apponit *mensam, sellam* locat, ingerit ultro
6.49 quos rebare *vagos, nexos* miraberis artus
6.296 comminus *excepit, mutavit* vera liquorem

γ²A

2.400 tigna *aras statuas* caementum saxa metallum
3.416 instructus *comis facilis* placabilis acer
4.188 irati instabiles *multi tumidi furiosi*
5.260 ungula *carnifices tortores* flagra catenae
6.397 frater mater *erus famulus* matrona propinquus

γ³B

6.188 atque *inopum sumptum* mox largitione levandum

δ¹A

1.93 ille etenim modico *contentos nos* iubet esse
2.211 quid simile *his titulis* tandem conferre valebit
3.130 auderet tantum *pro sancto* iure sacerdos
3.240 *qui, si finiri* culpam videt, abicit iram
3.395 quo clypeum regat *ingenio, quo* turbine telum
3.448 quin et fraternos *pro tanto* crimine fletus
4.57 dicite, *qui spreti* calcantes gaudia mundi
5.72 hos tantum adiunxit *socios, quos* culmen honoris
5.135 praesentis *domini, qui* pro virtute favoris.
5.152 quod *si pernici* propere translata volatu
5.187 ut *quae munificae* dederat documenta fidei
5.628 compulit *his dictis* raptim remeare retrorsum

δ¹B

1.245 saepe sacerdotes, *populum dum* saepe cohercet
1.328 irrupit *maestam tam* tristi funere cellam
1.368 nam *dum contiguum* Lupicini praeterit agrum
2.130 cellula. *tum solum* sociandi causa dabatur
2.171 martyris, et *quonam tam* clarae stemmata palmae
2.186 *tum dominum* tota fidei virtute precatus
2.568 *tum demum* motus tam iusta voce rogantis
2.725 *dum dominum* laudat, sibimet nil arrogat inde
3.37 si qua fuit regem *tumidum tum* causa rogandi

4.142	mox pestis collecta *locum. tum* vulnere solo
4.335	*tum demum* celeri linquens incendia saltu
4.639	pugnacemque *virum cum* pace ad bella remisit
5.701	*gemmarum, dum* sancta deo sollemnia defert
5.764	implentes *laceram tam* crebra aspergine puppem
6.228	effigians *illam quam* Christo iudice sumpsit
6.269	*quidam tam* properis caecato corde rebellans
6.481	et *tam contiguam* vicina pericula praedam
6.498	virtute interimit *flammam, quam* lumine nutrit

ε A

3.240	*qui, si finiri* culpam videt, abicit iram
3.253	quod *te de* mediis raperet clementia flammis
4.514	hic tantus labor, hic sumptus *quo, pro* dolor, exit

ζ A

1.223	cui *sanctus:* 'dominus defendit, non nocet hostis
2.53	sordida quod *vestis cordis* testata nitorem
2.117	*nulli vendendi* quicquam concessa facultas
2.154	ut doctrina *dei tali* de fonte fluentes
2.163	vana *superstitio falso* decepta sacrarat
2.197	hoc onus *adiecto duplicato* crescat honore
2.327	nam certe *nullus penitus* fuit agmine in illo
2.329	*adiectus gregibus* domini munimine signi
2.393	acrior ergo *animus penitus* virtute recepta
2.512	*ambitio blando* renuentem iure coegit
2.557	ac *pronus pedibus* blandas circumligat ulnas
3.18	prosperitas, *rebus gravius* ruitura secundis
4.318	angelus haec *domini mandati* verba profatur
4.326	quem postquam *solio subito* rex vidit ab alto
5.614	stellantis *solii tangi* per membra probatur
5.853	sed commissa *deo caelo* conclusit in alto
6.254	hospite tam *diro tanto* crudelior hostis

ζ B

2.35	excipit *egressum populorum* turba benignam
2.195	addere *supplicium, titulum* reddendo alienum
2.233	funeris *obsequium cultum* putat esse nefandi
2.349	sepsisse *implicitum flammarum* gurgite tectum
2.377	perfundens *totum lacrimarum* flumine vultum
3.113	anticipat *votum famulorum* cura vigorque
3.450	exorans *dominum lacrimarum* flumine Christum

4.427	permisit *fessum paulum* requiescere corpus
4.493	atque *utinam nostram* paulum rarescere noctem
4.603	verum ubi *lascivum rursum* rediviva voluptas
5.122	quidquid ad *ornatum commentum* dextra paravit
5.492	parcebat *paulum Turonorum* mitior urbi
5.571	reddit et *expressum nimborum* pugna fragorem
5.657	ille ubi *piscantum vanum* fluxisse laborem
5.684	obstipuere *suum monachorum* gaudia votum
5.715	*interitum, quorum* vitam poscebat ab ipso
5.716	compulsus *proprium paulum* laxare rigorem
5.755	instabiles *sursum nimium* tollentia colles
6.61	gurgitis et *cerebrum membrorum* mole perurguet
6.69	absorbet *tantum miserorum* membra profundum

PHOENIX (170 lines)

	A		B	
α^1	1	0.59	1	0.59
δ^1	1	0.59	–	
ϵ	1	0.59	–	

$$\alpha^1 \, A$$

94   *depositi tanti* nec timet illa fidem

$$\alpha^1 \, B$$

170   *aeternam vitam* mortis adepta bono

$$\delta^1 \, A$$

31   hoc nemus, *hos lucos* avis incolat unica Phoenix

$$\epsilon \, A$$

162   cui *de se* nasci praestitit ipse deus

PISCATORIS FURTUM

Anth. Lat. 8 Bailey (21 Riese)
(285 lines)

	A		B	
α^1	1	0.35	–	
α^5	1	–		
α^7	–		1	0.35
γ^1	1	0.35	–	
γ^2	2	0.70	–	
δ^2	1	0.35	–	
ζ	–		1	0.35

$$\alpha^1 \, A$$

5   pone *animos laetos,* quisquis testantia furtum

$$\alpha^5 \, A$$

273   dicite iam *poenas mandatas* legibus almis

$$\alpha^7 \, B$$

33   *contemptum, templum* pauper, piscator abundans

$$\gamma^1 \, A$$

200   auro ardet *Glauce, Danae* corrumpitur auro
201   auro emitur *Pluton, Phlegethon* transcenditur auro

$$\gamma^2 \, A$$

177   perditus *abiectus maledictus* sordidus amens

$$\delta^2 \, A$$

248   *hinc illinc* lucent collatis crimina rebus

$$\zeta \, B$$

158   ad causam *scelerum templum* patuisse rapinis

PRISCIANUS

	A		B			A		B	
De laude Anastasii (312 lines)					*Periegesis* (1087 lines)				
α^1	3	0.96	–		α^1	3	0.28	2	0.19
α^3	1	0.32	–		α^2	1	0.09	–	
β^2	1	0.32	–		α^3	2	0.19	–	
δ^1	7	2.24	–		α^4	2	0.19	–	
ζ	–		1	0.32	α^5	1	0.09	–	
					γ^2	2	0.19	1	0.09
					γ^3	1	0.09	–	
					δ^1	11	1.01	5	0.46
					ε	1	0.09	–	
					ζ	5	0.46	8	0.74

De laude Anastasii

α^1 A

49	gloria *magnanimi Traiani* cesserit isti
188	iam *patrias oras* tolerabant nave tenentes
238	praemia pro *poenis speratis* sumere mirans

α^3 A

| 190 | *disiectis claustris* quibus aequor frangitur album |

β^2 A

| 298 | *Hypatii vestri* referam fortissima facta |

δ^1 A

3	munere pro vitae, *pro pulchro* lumine solis
36	corpora nam, *spolii si* copia nulla dabatur
75	impetus hos frangit, *pulsos nos* demetit unguis
85	*pro merito* laudum cui nomen Isauricus illo
98	principe *pro iusto* taceat quis fulminis ictus
192	pro pietate *tua, qua* respicis omnia, fundunt
266	auxilio *summi, qui* conspicit omnia, patris

ζ B

| 294 | ore *canam quonam* pietatis culmina tantae |

Periegesis

α¹ A

443	*Haemus Threicius.* quem contra tenditur agris
454	Phocis ad *arctoos ventos* extenditur ingens
1023	*divitias magnas* sic tellus ista ministrat

α¹ B

270	ad *spatium multum terrarum* rura colentes
360	istuc *reginam propriam* venere secuti

α² A

53	*exiguos multos,* dirimens mare, quattuor amplos

α³ A

194	*extinctis populis* Nasamonum marte Latino
317	ignipotens *oleo commixto* gignitur herba

α⁴ A

662	cui sunt *vicini Sindi* quoque Cimmeriique
770	sunt *Chalcedonii primi* post ostia Ponti

α⁵ A

671	Aegypto *missi Colchi* tenuere coloni

γ² A

545	et celsa *Scyrus, Peparethus* sacraque Lemnos
637	in partes *euri zephyri* boreaeque vel austri

γ² B

104	*Tyrrhenum, Siculum* nec non simul Hadria vastus

γ³

250	haec *facies Libyes,* hic claudit terminus illam

δ¹ A

247	scilicet eoas *Nili qui* continet oras
383	cernitur hic tumulus, *qui Cadmi* dicitur esse
408	Alpheus, *qui Messeni* discinditur undis
447	campi, *quos medios* Acheloia persecat unda
647	Sauromatis *mixtae quae* Thermodonte relicto
734	Caspius *his populis* lustratur pontus et oris

763   Rhebas, *qui rapidi* decurrit ad ostia Ponti
800   ad mare vergentes; *qui Tauri* cornua cernunt
899   alterius *Syriae; quae* tendit adusque Sinopen
918   quod medias amnes *has terras* flumine cingunt
967   *terras, quas* Medi nemorosis montibus umbrant

δ^1 B

30   Caspia *quam mediam* faciunt atque aequora ponti
150   extra *quam, boream* quod scandit, sola Carambis
433   Thessalis et tellus *Macetum cum* moenibus altis
457   *quam mediam* dirimit descendens murmure vasto
533   nubila *quam numquam* vetuerunt lumine solis

ε A

1087   omnipotens *pro quo* genitor mihi praemia donet

ζ A

267   sic Europa *tibi facili* dinoscitur arte
503   hoc dederat *fluvius, cuius* generatur ad undas
658   aere *damnati tristi* nimioque rigore
765   nam prius *expositae Scythiae* sunt semina gentis
793   *Pactolus; cuius* prope ripas tempore verno

ζ B

217   non alius *tantum fluviorum* ditat harenas
242   ad cuius *zephyrum Macetum* sunt moenia regis
265   assimilant *conum laterum* compagibus aequis
270   ad *spatium multum terrarum* rura colentes
386   in senio, *postquam patriam* Thebasque relinquunt
789   ad latus *istarum boreum* sunt moenia clara
830   hoc esto *laterum boreum* tibi; Nilus ad oras
934   hoc quoque pingue *solum palmarum* divite silva

PRUDENTIUS

Ham., Psych., Symm. II
(3013 lines)

	A		B	
α^1	11	0.37	11	0.37
α^4	–		1	0.03
α^5	–		2	0.06
β^1	1	0.03	–	
γ^1	1	0.03	–	
γ^2	5	0.16	–	
γ^3	–		5	0.16
δ^1	15	0.49	11	0.37
δ^2	2	0.06	–	
ε	5	0.16	–	
ζ	5	0.16	8	0.26

α^1 A

Ham. 62	obvia *terrenis oculis* elementa, quibus se
Ham. 105	assignare *Deos proprios,* sua cuique iura
Ham. 143	*incautas animas* non cessat plectere, Nebroth
Ham. 389	his *aegras animas* morborum pestibus urget
Ham. 651	*obbrutescentis capitis,* ne pervia tales
Ham. 676	et super *ingenio proprio* laxaeque saluto
Ham. 893	abdita *corporeis oculis,* cum saepe quietis
Psych. 367	inde ad *nocturnas epulas,* ubi cantharus ingens
Psych. 607	causa *mali tanti* iacet interfecta; lucrandi
Psych. 670	inter *confertos cuneos* Concordia forte
Symm. 1094	spectatura *sacris oculis.* sedet illa verendis

α^1 B

Ham. 422	*postremum cuneum* rex promovet Euvaeorum
Ham. 519	frena *potestatum variarum* sustinet ac sub
Ham. 545	*ingenium, purum,* sapiens, subtile, serenum
Psych. 284	*extinctum Vitium* sancto Spes increpat ore
Psych. 415	implicat *excussam dominam: nam* prona sub axem
Symm. 69	allegat *morem veterem,* nil dulcius esse
Symm. 388	*summum consilium,* fida ut tutacula nudis
Symm. 424	*summorum procerum* fastigia, quos duodeni
Symm. 469	*antrum carcareum* dissolvite, corpora sub quo

| Symm. | 881 | et portenta *deum summum* numerosa putatis |
| Symm. | 1030 | aspera nam *segetem surgentem* vincula texunt |

$$\alpha^4\,B$$

| Symm. | 877 | haec putat esse *deum nullum,* namque omnia verti |

$$\alpha^5\,B$$

| Ham. | 960 | devoret hanc *animam mersam* fornacibus imis |
| Psych. | 338 | argento *albentem seriem,* quam summa rotarum |

$$\beta^1\,A$$

| Symm. | 447 | assignare *suos genios* perque omnia membra |

$$\gamma^1\,A$$

| Symm. | 529 | Aegypto *dederat, clangebat* bucina contra |

$$\gamma^2\,A$$

Ham.	73	subnixus ratione, *viget, splendet,* volat, auget
Ham.	349	una *voluntatis, iuris,* virtutis, amoris
Psych.	449	fibula, *flammeolum, strophium,* diadema, monile
Symm.	446	cum portis, domibus, *thermis, stabulis* soleatis
Symm.	558	*Fabricios, Curios,* hinc Drusos, inde Camillos

$$\gamma^3\,B$$

Ham.	115	*rerum principium* mortis de fomite traxit
Psych.	167	perque *cicatricum numerum* sudata recensens
Psych.	495	*terrarum vitium,* quod tantis cladibus aevum
Symm.	75	addit et *arcanum rerum* verique latebras
Symm.	796	*viventum meritum,* quos tantum pascere iussa est

$$\delta^1\,A$$

Ham.	204	fluxit origo *mali, qui* se corrumpere primum
Ham.	661	participes iubet esse *sui. qui, si* foret auctor
Psych.	377	his *vos imbutos* dapibus iam crapula turpis
Psych.	498	ausa sacerdotes *Domini, qui* proelia forte
Psych.	629	*his dictis* curae emotae, Metus et Labor et Vis
Psych.	753	barbaries, *sanctae quae* circumsaepserat urbis
Psych.	840	spiritus *his titulis* arcana recondita Mentis
Symm.	49	haec si non ita sunt, *edatur cur* sacra vobis
Symm.	443	Romam dico *viros, quos* mentem credimus urbis
Symm.	463	quae prohibet peccare *reos, quos* ferrea fata
Symm.	703	depulit *hos nimbos* equitum non pervigil anser

Symm. 721	*si potui* manibus Gallorum excisa levare
Symm. 754	aut candor perit *argenti si* defuit usus
Symm. 772	at non *a patria;* patriae sua gloria Christus
Symm. 953	indicio est annona, *tuae quae* publica plebi

$$\delta^1\,B$$

Ham. 23	distantes quoniam, *proprium dum* quisque revulso
Ham. 468	effudit Dominus, *populum dum* forte rebellem
Ham. 762	et *circum cum* plebe sua madidasque popinas
Psych. 79	quod fuerat, *Verbum, dum* carnis glutinat usum
Psych. 270	huc eques illa *dolum, dum* fertur praepete cursu
Psych. 296	dum tumet *indomitum, dum* formidabile fervet
Psych. 364	post immortalem *tunicam quam* pollice docto
Psych. 415	implicat *excussam dominam: nam* prona sub axem
Symm. 3	diximus, et nostro *Romam iam* credere Christo
Symm. 392	quem putet *auctorem, quem* rerum summa sequatur
Symm. 416	regius *exortam iam* tunc habuit status urbem

$$\delta^2\,A$$

Ham. 776	dividit *huc illuc* rapiens sua quemque libido
Psych. 192	*huc illuc* frendens obvertit terga, negata

$$\varepsilon\,A$$

Ham. 643	cum possit prohibere, sinat; *qui si* velit omnes
Ham. 661	participes iubet esse *sui; qui si* foret auctor
Ham. 682	dederet, arbitrium *de te* tibi credere avarus
Psych. 623	addubitas *ne te* tuus umquam deserat auctor
Symm. 153	ambitiosa velit, *ne se* popularibus auris

$$\zeta\,A$$

Ham. 293	pectitur. hunc *videas lascivas* praepete cursu
Ham. 892	expertus *dubitas animas* percurrere visu
Symm. 527	*difficilis operis* fuit immensique laboris
Symm. 579	*virtutis: dicis* domitum terraque marique
Symm. 1059	lexque *pudicitiae vitae* cum fine peracta

$$\zeta\,B$$

Ham. 110	ede *coheredum distinctum* ius dominorum
Ham. 752	attenuant *saxum, tantum* lambentibus umor
Ham. 918	haec ille ante *obitum membrorum* carcere saeptus
Symm. 33	*numquam pinnigeram* legio ferrata puellam
Symm. 176	*longaevam perimam* magico certamine matrem

Symm. 371	*sortitam quonam* genio proprium exigat aevum
Symm. 829	nec vitio *utentum restrictum* deficit aut se
Symm. 1003	cur non *Christicolum tantum* populatur agellos

REPOSIANUS (182 lines)

	A		B	
γ^2	–		1	0.55
δ^1	3	1.65	–	
ζ	1	0.55	–	

In 93 the reading of the Salmasianus *terribilem divum tuo solo* (ex *solum*) *numine victum* is obviously corrupt. My edition reads *tu solus numine vincis* (*tu … vincis* is Burman's).

γ^2 B

128   loricam *clipeum gladium* galeaeque minacis

δ^1 A

34   pictus amore deae; *si Phoebi* lumina desint
94   pro telis flores, *pro scuto* myrtea serta
113   in *flammas, quas* diva fovet. iam languida fessos

ζ A

8   quid conversa *Iovis laetaris* fulmina semper

RUTILIUS NAMATIANUS (712 lines)

	A		B	
α^1	1	0.14	1	0.14
γ^1	–		1	0.14
δ^1	7	1.00	–	
ζ	3	0.40	2	0.28

α^1 A

2.33   *excubiis Latiis* praetexuit Appenninum

α^1 B

1.7   qui *Romanorum procerum* generosa propago

γ^1 B

2.59   hic *immortalem, mortalem* perculit ille

δ^1 A

1.6   *nasci felici qui* meruere solo
1.57   volvitur ipse *tibi, qui* continet omnia, Phoebus
1.158   *si colui* sanctos consuluique patres
1.165   *his dictis* iter arripimus: comitantur amici
1.217   solvimus Aurorae *dubio, quo* tempore primum
1.230   *qui pastorali* cornua fronte gerit
1.601   nec mirum, *magni si* redditus indole nati

ζ A

1.6   *nasci felici qui* meruere solo
1.279   paulisper *litus fugimus* Munione vadosum
1.303   insidias *paci moliri* tertius ausus

ζ B

1.264   *auctorem pecudem* fons Heliconis habet
1.395   atque *utinam numquam* Iudaea subacta fuisset

SEDULIUS

Carmen Paschale (1753 lines)

	A		B	
α^1	1	0.06	–	
α^5	1	0.06	–	
β^1	2	0.11	–	
γ^1	–		1	0.06
γ^2	–		1	0.06
δ^1	13	0.74	5	0.28
ε	2	0.11	–	
ζ	5	0.28	5	0.28

α^1 A

3.264   *pisciculis paucis* et septem panibus agmen

α⁵ A

$$\alpha^5 \, A$$

5.208   huic reus est *mundus, salvatus* sanguine iusto

$$\beta^1 \, A$$

4.81   munda *suis lacrimis* redit et detersa capillis
4.159   nulla *meis famulis* feritas adversa nocebit

$$\gamma^1 \, B$$

3.252   millia *caecorum, claudorum* millia passim

$$\gamma^2 \, B$$

1.34   semper *principium, sceptrum* iuge, gloria consors

$$\delta^1 \, A$$

1.61   *qui coeli* fabricator ades. qui conditor orbis
1.242   heu *miseri, qui* vana colunt, qui corde sinistro
1.305   demens, *perpetui qui* non imitanda parentis
1.327   impugnant sua dicta *viri, qui* brachia nudis
2.269   debita *laxari qui* nobis cuncta rogamus
3.21   more *Dei, qui* cuncta prius quam nata videndo
3.190   ille chelydrus adest, *nigri qui* felle veneni
5.48   sufficerent magni *fuso pro* sanguine Christi
5.49   qui pater est *mundi, qui* fecerat, hunc quoque nasci
5.75   plus duodena *dari, si* mallet sumere poenas
5.101   namque per *hos colaphos* caput est sanabile nostrum
5.103   *his alapis nobis* libertas maxima plausit
5.158   crimina iudicio, *vigili si* mente notares

$$\delta^1 \, B$$

1.566   gaudia longa *metam: nam* qui deflemus in Adam
5.4   non aliam, sed rursus *eam quam* munere plenam
5.54   sors melior nescire *datam, quam* perdere vitam
5.178   quod *vinum cum* felle datum, tristemque saporem
5.258   sicut in *horrendum dum* convertuntur acetum

$$\varepsilon \, A$$

2.205   nam scriptura decet *de te* mandasse Tonantem
4.5   nil igitur summo *de se* sperantibus umquam

$$\zeta \, A$$

1.66   qui *stellas numeras,* quarum, tu nomina solus
3.165   sumpsistis *gratis, cunctis* impendite gratis

4.229   orat *inexhausti tribui* sibi dona fluenti
4.243   nec poterat *quisquam festucam* vellere parvam
5.103   *his alapis nobis* libertas maxima plausit

ζB

2.196   Christus ad haec: '*tantum Dominum* scriptura Deumque
2.275   denarios *centum conservum* senserit ullum
4.127   et grege *discipulum, miserum* cum cominus ecce
5.33   prodidit *auctorem, panem* cui tradidit ipse
5.170   implet harundo *manum, sceptrum* quod mobile semper

SERENUS

Liber medicinalis (1107 lines)

	A		B	
α^1	1	0.09	–	
δ^1	5	0.45	1	0.09

The only alphas in our texts are 82 and 502 *qui subito raptos ventos sucosque revellit*, but the latter verse is corrupt in the principal manuscript and omitted in the others. The omega count is sensationally low; in 770 *fas erit et tristem saltem mulcere dolorem* we may as well write *saltim*, as in editions before Baehrens.

α^1

82   aut *acido Baccho* miscebis farra lupini

δ^1A

586   *quos ternos* tepida mixtos hausisse medella est
651   quod *si feminei* properabit sanguinis imber
853   toxica *praeterea qua* sint pellenda medella
1032   *qui primi* fuerint pullo crescente caduci
1038   *qui veteri* claras expressit more togatas

δ^1B

139   cerussam et *cartam quam* gens Aegyptia mittit

SIDONIUS
I–VII (1847 lines)

	A		B	
α^1	6	0.33	2	0.11
α^3	3	0.16	–	
α^5	–		1	0.05
β^1	1	0.05	–	
β^2	1	0.05	–	
γ^1	2	0.11	–	
γ^2	10	0.54	1	0.05
γ^3	–		4	0.22
δ^1	13	0.70	12	0.65
ε	3	0.16	–	
ζ	9	0.49	2	0.11

α^1 A

2.364	*Vandalicas turmas* et iuncti Martis Halanos
5.204	exclusa *sceptris Geticis,* respublica si me
6.8	*armatus partus* vertice dividuo
7.32	*Saturnus profugus,* vaga Cynthia, Phoebus ephebus
7.33	Pan pavidus, *Fauni rigidi, Satyri* petulantes
7.342	et *populis Geticis* sola est tua gratia limes

α^1 B

| 2.108 | protulit *undantem segetem* sine semine campus |
| 7.372 | Francus *Germanum primum* Belgamque secundum |

α^3 A

2.274	*metato spatio* castrorum Serdica vidit
5.58	hic praedo et *dominis exstinctis* barbara dudum
5.489	hic tu vix *armis positis* iterum arma retractas

α^5 B

| 5.524 | *instantem iuvenem;* quisquis fortissimus ille est |

β^1 A

| 4.14 | atque *meae vitae* laus tua sit pretium |

β^2 A

| 2.232 | hic primum ut *vestras aquilas* provincia vidit |

γ^1 A

5.319	ut *titulis, meritis* fuerat. res ordine currit
7.105	et caligas *Gai Claudi* censura secuta est

γ^2 A

2.171	*deflevit risit tacuit;* quodcumque Platonis
2.176	quidquid *Anaximenes Euclides* Archyta Zenon
5.336	*Gaetulis Nomadis* Garamantibus Autololisque
5.475	*Pannonius Neurus Chunus* Geta Dacus Halanus
5.476	*Bellonotus Rugus* Burgundio Vesus Alites
5.588	nectet *muralis vallaris* civica laurus
7.80	Sulla Asiatogenes *Curius Paulus* Pompeius
7.81	Tigrani *Antiocho Pyrrho* Persae Mithridati
7.236	cursu *Herulus Chunus* iaculis Francusque natatu
7.323	*Chunus Bellonotus Neurus* Bastarna Toringus

γ^2 B

2.415	*costum malobathrum* myrrhas opobalsama tura

γ^3 B

2.318	*divorum numerum.* quem mox Oenotria casum
7.202	quid *volucrum studium,* dat quas natura rapaces
7.450	*exsilium patrum,* plebis mala, principe caeso
7.458	postquam in *consilium seniorum* venit honora

δ^1 A

2.191	*quo genio* Plautus, quo fulmine Quintilianus
2.192	*qua pompa* Tacitus nunquam sine laude loquendus
2.369	illius esse viri *quo viso,* Vandale, semper
2.394	applicat *a laeva* surgentem balteus ensem
2.454	sola *tenes: res* empta mihi est de sanguine Crassi
2.496	Hippomanis iam non *pro solo* colligat auro
2.502	*hos thalamos,* Ricimer, Virtus tibi pronuba poscit
5.284	texta Nabataeum *pro Rheno* sulcat Hydaspen
5.326	*his titulis* princeps lectus similique labore
7.24	prime venis, *viridi qui* Dorida findere curru
7.67	Samnitem, Gurges, *Volsci qui* terga cecidit
7.279	proditus ut tandem *tanti qui* causa tumultus
7.549	non latet: *his tantis* tibi cessimus, inclite, bellis

δ^1 B

1.20	quamvis *hinnitum, dum* canit, ille daret
2.81	*tum demum* Babylon nimis est sibi visa patere

2.338	sulcabat *madidam iam* torrens alveus alvum
4.5	sed rus *concessum dum* largo in principe laudat
5.108	Illyricum rexisse *solum cum* tractibus Histri
5.391	Vandalus opperiens *praedam, quam* iusserat illuc
5.436	*praedonum tum* forte ducis, cui regis avari
6.1	Pallados armisonae *festum dum* cantibus ortum
6.3	et *dum Mopsopium* stipantur per Marathonem
7.210	vectigal. *procerum tum* forte potentior illic
7.251	huius *tum famulum* quidam truculentior horum
7.530	gaudens turba *virum. procerum tum* maximus unus

ε A

2.17	ne tempestates, *ne te,* pirata, timeret
2.26	*de te* non totum licuit tibi. facta priorum
7.506	quod te, Roma, capit; sed *di si* vota secundant

ζ A

2.94	huic socer *Anthemius, praefectus,* consul et idem
5.299	emeritus *iuvenis, sterilis* ieiunia terrae
5.310	non animum *populi): ferri* mala crimina ferro
5.463	dite magis *ferro, merito* cui subiacet aurum
7.58	fundamenta *iugis aperis* mihi, Romule pauper
7.107	Galbam sternis, *Otho, speculo* qui pulcher haberi
7.138	fessa modo *possis, paucis,* cognosce, docebo
7.402	mole *magisterii legati* iura subisse
7.438	Romulus et *Tatius foedus* fecere, parentum

ζ B

2.312	vulnera *Tantalidum, quorum* tibi funera servat
7.530	gaudens turba *virum. procerum tum* maximus unus

SYMPOSIUS (317 lines)

	A		B	
α^1	7	2.20	1	0.31
α^2	2	0.60	–	
α^4	1	0.31	–	
α^7	3	0.94	–	
γ^1	2	0.60	–	
γ^2	–		1	0.31
δ^1	2	0.60	1	0.31
ζ	1	0.31	1	0.31

α^1 A

5	post *epulas laetas*, post dulcia pocula mensae
46	nunc *rigidi caeli* duris connexa catenis
115	*sanguineas praedas* quaerens victusque cruentos
200	scrutor *aquas medias, ipsas* quoque mordeo terras
202	et manet in *mediis undis* immobile robur
226	ius *avidae linguae*, finis sine fine loquendi
253	iuncta *solo plano*, manibus compressa duabus

α^1 B

300	inter *luciferum caelum* terrasque iacentes

α^2 A

56	curro *vias multas* vestigia nulla relinquens
306	*insidias nullas* vereor de fraude latenti

α^4 A

136	pes *unus solus*, sed pes longissimus unus

α^7 A

161	laetus honor *frondis, tristis* sed imago doloris
200	scrutor *aquas medias, ipsas* quoque mordeo terras
242	infundor *lymphis, gelidis* accendor ab undis

γ^1 A

75	tarda gradu *lento, specioso* praedita dorso
126	dissimilis *matri, patri* diversa figura

γ^2 B

175	nolo *virum thalamum:* per me mea nata propago est

δ^1 A

60	non possum *nasci, si* non occidero matrem
140	et tamen *hos ipsos* omnes ego porto supinos

δ^1 B

58	nondum natus eram, nec *eram iam* matris in alvo

ζ A

42	ex *alto venio* longa delapsa ruina

ζ B

69	porto *domum mecum*, semper migrare parata

MEDIEVAL

	α^1	α^{1-6}	ω
Alcuin (1657)	1.02 (0.72 + 0.30)	1.14 (0.84 + 0.30)	2.71 (2.35 + 0.36)
Angilbert (758)	1.85 (1.19 + 0.66)	2.10 (1.44 + 0.66)	1.72 (0.92 + 0.80)
Grosseteste (1476)	0.61 (0.27 + 0.34)	1.30 (0.68 + 0.62)	1.96 (1.15 + 0.81)
Hrotsvitha			
Maria (903)	1.21 (0.77 + 0.44)	2.10 (1.32 + 0.78)	3.76 (3.10 + 0.66)
Pelagius (412)	0.97 (0.97 + 0)	1.45 (1.21 + 0.24)	1.94 (1.70 + 0.24)
Theophilus (455)	1.76 (1.54 + 0.22)	2.86 (2.42 + 0.44)	4.40 (3.74 + 0.66)
Hugo of Liège (2001)	1.00 (0.85 + 0.15)	1.75 (1.40 + 0.35)	1.60 (1.25 + 0.35)
Matthew of Vendôme			
Epistule (1808)	0.49 (0.22 + 0.27)	0.71 (0.44 + 0.27)	2.21 (1.77 + 0.44)
Tobias (2226)	0.67 (0.45 + 0.22)	0.99 (0.72 + 0.27)	2.02 (1.48 + 0.54)
Nigel de Longchamps			
(2690)	0.48 (0.33 + 0.15)	0.69 (0.43 + 0.26)	1.82 (1.23 + 0.55)
Ps.-Ovidius (818)	2.32 (2.08 + 0.24)	3.78 (3.42 + 0.36)	3.54 (2.20 + 1.34)
Theodulf (1716)	1.05 (0.82 + 0.23)	1.10 (0.87 + 0.23)	1.28 (1.05 + 0.23)
Marcus Valerius (451)	0.22 (0.22 + 0)	0.22 (0.22 + 0)	2.19 (1.75 + 0.44)
Waltharius (1496)	0.42 (0.42 + 0)	0.69 (0.62 + 0.07)	2.90 (1.38 + 1.52)
Ysingrimus (3994)	0.26 (0.18 + 0.08)	0.40 (0.30 + 0.10)	1.31 (0.88 + 0.43)

In a total 22861 lines I find 175 α^1s (0.73%), 81 α^{2-6}s (0.35%), roughly half as much again as compared with the figures for religious versifiers of late antiquity, to say nothing of classics and post-classical classicizers. Omegas also average higher 474 (= 2.08%). The pastorals of Marcus Valerius, which approach closest to classical standards, are notably low in alphas (but only 451 lines to go on), though not much lower than the semibarbarous Ysingrimus. Stylistic factors, such as I suggest in Matthew of Vendôme's case, may account for something, but the general trend is plain. The medieval Latin versifier took homs in his stride.

ALCUIN

De sanctis Euboricensis ecclesiae
(1657 lines)

	A		B	
α^1	12	0.72	5	0.30
α^2	2	0.12	–	
α^7	3	0.18	–	
β^1	5	0.30	2	0.12
β^2	1	0.06	–	
γ^1	2	0.12	–	
γ^2	8	0.48	1	0.12
γ^3	1	0.12	–	
δ^1	17	1.03	2	0.12
ε	7	0.43	–	
ζ	17	1.03	4	0.24

α^1 A

35	quo *variis populis* et regnis undique lecti
359	quae fuit *Oswaldi sancti* iam filia fratris
364	ossa *viri sancti* summi ad fastigia caeli
367	nam *priscis odiis* assumere primitus ossa
461	*aequoreos populos* lata quae clade peremit
502	*divisis linguis populis* per nomina patrum
719	*inmissis lymphis* quendam sanaverat aegrum
779	atque *aegris oculis* praestaverat ipsa medelam
1294	cui iam *praeclarus Ceolfridus* praefuit abbas
1448	nam rudis et *veteris legis* patefecit abyssum
1506	ast nova *basilicae mirae* structura diebus
1543	quidquid *Gregorius summus* docet et Leo papa

α^1 B

103	insuper *imperium latum* tibi terminat undis
350	pulvere *sacratam postam* timet igneus ardor
696	*cernentem fratrem* morbo culpaque relaxat
1131	inque *modum mirum* toto de corpore tota
1600	cui quoque *praesentem testem* me contigit esse

α^2 A

1228	*ecclesias alias* donis ornavit opimis
1564	post *annos binos*, menses simul atque quot annos

α⁷ A

240	millibus *innumeris, spoliis* nimiumque superbum
1374	ambulat ergo *freto solido* ceu tramite terrae
1522	tradidit *Eanbaldo dilecto* laetus alumno

β¹ A

30	hanc piscosa *suis undis* interluit Usa
477	nec tamen esse *mei meriti* scio vivere tantum
1060	millia et usque *suos socios* undena natabant
1468	officiumque *suis meritis* decoraverat almis
1599	atque *mei pueri* causam succinge parumper

β¹ B

102	ecce *tuam vitam* quaerenti servat ab hoste
496	ad dominumque *suam vitam* converterat omnem

β² A

1385	sic precibus *nostras animas* evadere fluctus

γ¹ A

269	*pauperibus largus, parcus* sibi, dives in omnes
1608	fulgidus *aspectu, statu* sublimis honesto

γ² A

123	Saxonum *populus, Pictus* Scotusque, Britannus
502	*divisis linguis populis* per nomina patrum
621	*discipuli socii* stabant hinc inde gementes
1397	vir bonus et *iustus, largus,* pius atque benignus
1399	ecclesiae *rector ductor* defensor alumnus
1450	hos sibi *coniunxit docuit* nutrivit amavit
1553	quae Maro *Virgilius, Statius* Lucanus et auctor
1556	Servius *Euticius Pompeius* Comminianus

γ² B

1444	naturas *hominum pecudum* volucrumque ferarum

γ³ A

676	quaeque *dei famuli* sacrata est insula morte

δ¹ A

10	victrices aquilas *caeli qui* fertis in arcem
23	sustinuit merito, *mundi qui* sceptra regebant

100	rex deus aeternus, *caeli qui* sydera fecit
107	nuntius *his dictis* subito discessit ab illo
172	annuit *his dictis* senior, paucisque respondit
356	claruit *his signis* postquam locus iste peractis
366	maior *relliquias quas* tunc aulea tegebat
537	*his actis* etiam hostiles ut vidit ubique
985	omnia quae *vidi, si* scirem forte, requirit
1037	plurima *qui populi* Fresonum milia Christi
1063	lumen et hoc *illi, qui* sanctos ante necabant
1164	atque preces *domino pro* huius fundere vita
1358	cum lacrimis *domino pro* culpa supplicat illa
1407	de *quo versifico paulo* plus pergere gressu
1509	*suppositae quae* stant curvatis arcubus, intus
1526	librorum nato, *patri qui* semper adhaesit
1613	*his dictis* subito nitidus disparuit hospes

$$\delta^1 \, B$$

291	tempore *nam quodam* praesul sanctissimus Aedan
1292	utpote *septennem quem* fecit cura parentum

$$\varepsilon \, A$$

6	ut mea lingua queat *de te* tua dicere dona
440	denique praecisae *de te* ducuntur ubique
713	*de se* deque aliis, quae praescius ante videbat
797	cui comes instituit *de se* narrare, quis esset
905	'splendidus' inquit 'erat, qui *me de* corpore duxit
1162	*pro quo* pontificem lacrimans dux ipse rogabat
1182	ex sociis, ludo *ne se* misceret inani

$$\zeta \, A$$

91	Euborica *genitus, dominus* per cuncta futurus
183	*terribilis qualis* curvo fit Parthus in arcu
269	*pauperibus largus, parcus* sibi, dives in omnes
450	nescius in *gremio somno* praeventus habebat
452	ecce aliquid *gelidi lateri* sibi sensit adesse
535	milite cum *raro, primo* sed numen Olympi
624	atque levans *oculos, socios* conspexit et infit
918	ductor et ille *mihi meditanti* talia dixit
942	ad *poenas animas* ululantes quasque trahentes
1018	*pauperibus largus,* sibimet sed semper egenus
1082	*appositus patribus* suprema sorte quievit
1111	et pleno *penitus mutus* sermone locutus

1152	reddidit atque *deo proprio* cum coniuge grates
1200	ergo *gravi veluti* de somno surgeret ille
1342	quae esset, cur *fugeret, faceret* vel quae mala? cui tunc
1407	de *quo versifico paulo* plus pergere gressu
1574	pontificis *summi, nostri* patris atque magistri

ζ B

349	consumpsere *domum, nimium* res mira sed acta est
428	crus veneranda *canam, nequeam* si promere dignos
1420	ingenio *tantum librorum* proficiebat
1631	vidisti *sedem. iuvenem* nec sermo fefellit

ANGILBERT

Carmina (758 lines)

	A		B	
α^1	9	1.19	5	0.66
α^2	2	0.26	–	
α^5	–		1	0.13
α^7	2	0.26	1	0.13
β^1	2	0.26	1	0.13
γ^1	2	0.13	–	
γ^2	3	0.39	–	
γ^3	–		2	0.26
δ^1	3	0.39	–	
δ^2	1	0.13	–	
ε	1	0.13	–	
ζ	2	0.06	3	0.39

α^1 A

6.106	hic alii *thermas calidas* reperire laborant
6.208	per *patulas portas;* certatim exire senatus
6.427	altus et in *nudo campo* iacet, undique largo
6.471	et *latos clipeos,* galeasque et spicula; peltae
6.522	ex hinc *officiis divinis* rite peractis
6.531	post *laetas epulas* et dulcia pocula Bachi
7.44	perque *cruoriferos umbos,* per tela duelli
7.49	*raucisonos tinctos* furva nigredine corvos
7.51	*alipedes griphes* subito harpeiasque volucres

$$\alpha^1\,B$$

6.315	*frondosum lucum*, patulis fontesque recentes
6.321	*maturum populum natum* melioribus annis
3.330	*truncatam linguam* horrendaque multa gerentem
6.335	miraturque; *piam curam* gerit ille fidelem
7.10	*pompiferum mundum*, dura sub morte iacentem

$$\alpha^2\,A$$

1.38	ut *paucis verbis* plurima vota feras
7.23	iam *septingentos finitos* circiter annos

$$\alpha^5\,B$$

6.321	*maturum populum natum* melioribus annis

$$\alpha^7\,A$$

6.428	vestitus *spatio; celso* de colle videri
7.57	sub patris et *geniti, sancti* sub fluminis albi

$$\alpha^7\,B$$

6.99	hic iubet esse *forum, sanctum* quoque iure senatum

$$\beta^1\,A$$

6.14	sol nitet ecce *suis radiis:* sic denique David
6.74	atque *suis dictis* facundus cedit Homerus

$$\beta^1\,B$$

3.12	atque *tuum famulum* redde *tuo domino*

$$\gamma^1\,A$$

6.118	saxa: alii *subeunt, volvunt* ad moenia rupes

$$\gamma^2\,A$$

1.58	quidve *duces comites*, quid puer atque senes
6.63	mitis *praecipuus iustus* pius inclitus heros
6.66	*pacificus largus* solers hilarisque venustus

$$\gamma^3\,B$$

6.340	litora, *Saxonum populum* domitare rebellem
7.69	*cunctorum meritum* trilibri tunc lance librando

δ¹A

1.45	sic *vos coniunctos* defensio diva per annos
6.201	restaurat *proprii qui* publica gesta parentis
6.393	eripite *his terris,* David, me obtutibus almis

δ²A

6.146	*huc illuc* timido discurrit dammula gressu

εA

3.11	perfice quod *de me* coepisti mente benigna

ζA

3.14	*Elegius, precibus* auxiliare tuis
6.468	*pontifici celeri* cursu occurremus opimo

ζB

6.194	gaudet equo, et *iuvenum circum* manus emicat ardens
6.363	unius in *casum multorum* saevit hiatus
6.452	pastor apostolicus *centum cum* milibus altum

GROSSETESTE

De mundo et partibus (1476 lines)

	A		B	
α^1	4	0.27	5	0.34
α^2	4	0.27	1	0.07
α^4	2	0.13	3	0.20
α^7	2	0.13	1	0.07
γ^1	3	0.20	–	
γ^2	2	0.13	–	
γ^3	1	0.07	3	0.20
δ^1	9	0.62	5	0.34
ε	1	0.07	–	
ζ	5	0.34	3	0.20

α¹A

395	hec de *divinis scripturis* colligo scripta
403	in *vero caelo* nil mobilitatis habetur

729   corruit ergo *chorus sceleratus* de paradiso
1023   signifer *obliquo spacio* complectitur ambos

α^1 B

301   hinc deus hunc *mundum pulcrum solidum* sine terra
446   tres *multum spacium,* plus tribus ignis habet
466   *rerum simplicium* corpora nomen habent
707   cum Deus *angelicam naturam* conderet olim
1268   hinc *signum plenum* constat habere duas

α^2 A

872   sub *cancro solo* sedit in ede soror
1017   sub *signis istis* errancia sidera currunt
1267   ortus *dimidii signi* plenam facit horam
1305   non *horas senas* bissexti materiales

α^2 B

1387   *nullarum rerum* voces tonitrusque fragores

α^4 A

897   *quintus Mercurius, sextus* Sol, septima Luna
1403   *estates similes* de racione forent

α^4 B

301(?)   hinc Deus hunc *mundum pulcrum solidum* sine terra
319   *mundum perpetuum* facit hec concordia discors
1225   ni *cignum nigrum corvum* mihi feceris album

α^7 A

863   primus in orbe *deus, primus* pater ille deorum
1019   Sol cum *Saturno medio* meat ordine; Lunam

α^7 B

1225   ni *cignum nigrum corvum* mihi feceris album

γ^1 A

407   ergo – *velis, nolis* – duo sunt caeli, nec in uno
929   si *taceo, perdo* studium perdisque laborem
1036   blanda *bonis, duris* mollia certa iocis

γ^2 A

224   affectat *credit scit* sapit optat amat
467   *mobilior levior* subtilior omnibus ignis

γ³ A

1289   anni *tocius tempus* placet in duodenos

γ³ B

571   *sensum membrorum* sine membris possidet, omnes
573   *spirituum numerum* solus, qui condidit illos
975   maxima *zonarum spacium* complexa duarum

δ¹ A

233   *philosophari si* bene vis, mathematicus esto
249   has athomos, *has particulas* elementa vocamus
251   scire volo *si sentiri* simplex elementum
948   ludere *pro metro, pro* levitate metri
963   rursus et *hii tropici* dicuntur solsticiales
1204   eclipsis *Lune me* reparare potest
1238   *que stelle* tales et mutans regna cometes
1323   unum Luna diem *qua causa* iugiter annos
1419   si Saturnus in *his signis* cum sole moretur

δ¹ B

803   intentoque nimis *dum sextum* lumine lustro
1029   mundi spera *suum dum* circumvolvitur axem
1131   nunc monoides est, *quam 'primam'* dicimus; ex tunc
1269   ergo zodiacus, *dum totum* vertitur, horas
1400   Solis, *dum sensum* micior aura fovet

ε A

399   non tamen omnino tibi nolo credere: *ne me*

ζ A

713   hos, qui de *seraphin cherubin* thronis ceciderunt
897   *quintus Mercurius, sextus* Sol, septima Luna
914   quicquid *heri didici,* dedocet ista dies
1179   haec *terra umbra* quanto productior exit
1183   set *memini scripti* 'labor improbus omnia vincit'

ζ B

195   dum bene *quesiero, pacto* cogeris ad omne
1217   non *aliam reddam,* quia non opus est, racionem
1271   quisque *planetarum, quantum* sub quoque moratur

HROTSVITHA

	A		B	
Maria (903 lines)				
α^1	7	0.77	4	0.44
α^2	5	0.55	–	
α^3	1	0.22	–	
α^5	1	0.11	–	
α^7	2	0.22	2	0.22
β^1	2	0.22	2	0.22
γ^2	–		1	0.11
γ^3	1	0.11	4	0.44
δ^1	10	1.11	1	0.11
ε	4	0.44	–	
ζ	15	1.66	3	0.33
Pelagius (412 lines)				
α^1	4	0.97	–	
α^3	1	0.24	–	
α^7	1	0.24	–	
δ^1	4	0.97	–	
ζ	3	0.72	1	0.24
Theophilus (455 lines)				
α^1	7	1.54	1	0.22
α^2	3	0.66	–	
α^3	1	0.22	–	
α^4	–		1	0.22
α^7	1	0.22	–	
β^1	4	0.88	–	
γ^2	1	0.22	1	0.22
δ^1	11	2.42	1	0.22
ε	3	0.66	–	
ζ	5	1.10	2	0.44

Maria

α^1 A

73   ut *propria substantiola* bene multiplicata
78   volveret in *summa fortuna* lustra rotata
159   dic rogo, *divinae causae* quid pertinet ad me

287   quem merito *cives caelestes* laude frequentant
445   cui mox *iudiciis divinis* ista iubentur
632   de *parvo puero* praesepibus inveniendo
813   talibus ac *tantis signis* iam sedulo visis

$$\alpha^1 B$$

259   ecce *virum proprium* praesentem sentio salvum
612   moreque Iudaico *proprium meritum* recitando
834   atque *deum verum* magna virtute deorum
853   iure *deum verum* contestantur fore solum

$$\alpha^2 A$$

181   concipiet natam *cunctis saeclis* venerandam
266   progenuit natam *cunctis saeclis* venerandam
576   gaudens divinam *cunctis saeclis* venerandam
678   odis ex *ipso iusto* Simeone vetusto
682   nam post haec *binis mansurnis* saepe repletis

$$\alpha^3 A$$

755   et rursum *versis claris* aspexit ocellis

$$\alpha^5 A$$

618   pauperibus *cunctis, lectis* de finibus orbis

$$\alpha^7 A$$

114   legali *domino, devoto* legis amico
241   qui gregibus *lectis, silvis* discessit ab illis

$$\alpha^7 B$$

275   mandans *egregiam Mariam* vocitare puellam
704   iuxta *speluncam sanctam* cum prole Mariam

$$\beta^1 A$$

17   tu dignare *tuae famulae* clementer adesse
793   *de te* quippe *meo praecepto*, palmula, mando

$$\beta^1 B$$

234   cumque *suum dominum* terra videre locatum
887   et sentire *tuum solidum* per tempora regnum

$$\gamma^2 B$$

419   atque *sacerdotum vatum* pariterque priorum

γ^3 A

503	creditur ad velum *domini templi* pretiosum

γ^3 B

289	*ascensum graduum* subito ter quinque supinum
309	*ascensum graduum* cunctis patefecit in aevum
412	*ascensum graduum* conscendit namque superbum
626	tangendo *minimum pannorum* denique filum

δ^1 A

31	olim sed stultum *fari qui* iussit asellum
42	quos piget altithrono psallere *pro modulo*
207	angelus *his votis,* ut iussit, rite peractis
231	ipsius ⟨a⟩ *sexta,* ni fallor, forte diei
452	nam sibi *pro signo* dictum fuerat manifesto
564	*his dictis* Mariam blande conversus ad almam
579	se *pro salvando* venturum praescia mundo
616	tu *scis, praeceptis* fueram quod sedula legis
757	*his visis* lingua formavit talia verba
815	utitur *his verbis,* submisso munere fusis

δ^1 B

6	*vitam, quam* virgo perdiderat vetula

ε A

619	qui maerens venit, *de me* quoque laetior ivit
776	ut quantum libeat, *de te* mater mea carpat
793	*de te* quippe *meo praecepto,* palmula, mando
800	*de te* vincenti dicetur protinus illi

ζ A

122	his ita *finitis, sublatis* cernit ocellis
133	causa *iudicii iussisti* denique recti
140	angelus *astrigero subito* descendit ab alto
172	hortaris *regredi subdi* primoque pudori
179	aequa ferens *veni permagni* gaudia doni
201	quapropter *moneo, domino* libamine sacro
294	astantes *populi templi* pariterque ministri
379	pontifices *templi pretii* non munere parci
486	mitti cum *Maria causa* solaminis ergo
545	Caesaris Augusti censumque *sibi profiteri*
675	sistitur in *proprio parvo* cum munere templo

725   *quamvis humanis* sim parvus homuncio membris
791   cumque *profecturi deserti* per loca vasti
801   ad palmam *magni venisti* namque triumphi
811   implevit, magno nosmet *studio satiando*

ζ B

302   nec *mirum, sursum* coepit si figere gressum
631   visio *pastorum, signum* sibimet quoque dictum
766   si saltim *valeam puram* conprendere lympham

Theophilus

α¹ A

23   ac *castis viduis* nec non *cunctis peregrinis*
47   donec invitus trahitur *turbis glomeratis*
60   atque Theophilum *summis meritis* venerandum
171   *mundanae pompae* vano seductus amore
324   post haec *attonitis oculis* cernentibus ipsis
331   impetraque tuo veniam *famulo sceleroso*
353   sed nec *tartareis poenis* umquam capieris

α¹ B

303   huncque *deum verum* necnon hominem fore plenum

α² A

23   ac *castis viduis* nec non *cunctis peregrinis*
163   heu mihimet misero, *cunctis probris* vitiato
173   tempore iudicii *sanctis ipsis* metuendi

α³ A

310   postremo, *sacris expansis* in cruce palmis

α⁴ B

265   ceu numquam *visum carum* negat ergo magistrum

α⁷ A

31   quo nam *defuncto gremio* terraeque locato

β¹ A

37   hoc quoque persuadere *suo metropolitano*
327   proque *suis meritis* reddet bona vel mala cunctis
352   atque *tuae lacrimae* scelerum veniam meruere
445   plasma *suae dextrae* rapiens serpentis ab ore

$$\gamma^2 A$$

305 obprobiis tangi *colaphis, alapis* quoque caedi

$$\gamma^2 B$$

405 laudandam *mitem dulcem* Christi pietatem

$$\delta^1 A$$

216 dic rogo, *quis oculis* possim mis cernere prolis
249 Iudaici, plebem *domini qui* rexit herilem
253 complexus carae *licito quo* posset habere
284 *quo pacto, quo* iure quidem contingere tandem
338 emptum, *pro mundo* qui fusus erat perituro
344 *his dictis,* subito discessit virgo sacrata
360 *spero, quo* poenis pabulum non tradar amaris
375 *his dictis, oculis* iterum vigilabat apertis
380 *qua visa* membris mox contremuit resolutis
395 intonat *his verbis* mirantis voce profusis
421 *his dictis,* cartam comburebat maledictam

$$\delta^1 B$$

439 in quo nactus erat *veniam, quam* flendo rogabat

$$\varepsilon A$$

240 despiciendo deum *de te* sine sorde profusum
301 ut *de te* casta necnon de flamine sancto
417 ac *de te* genitum regem dominumque polorum

$$\zeta A$$

169 qui miser *elegi subdi* Satanae ditioni
218 quove *modo solio* praesens astare tremendo
226 invigilare *meo cerno* saepissime templo
270 talibus ac *tantis, aliis* multisque figuris
375 his *dictis, oculis* iterum vigilabat apertis

$$\zeta B$$

191 *eiusdem celerem* supplex quaero pietatem
244 post *lapsum scelerum* veniam meruere suorum

Pelagius

$$\alpha^1$$

207 hos et *amicitiae propriae* coniungere velle
323 fluctivagosque greges *variis laqueis* capientes

365   *aequa gloriola* regnarent mortua membra
398   iuxta naturam *fragilis carnis* perituram

$$\alpha^3 A$$

31   ac *reliquo victo* tanta de caede popello

$$\alpha^7 A$$

302   iudicis et *veri, caeli* super astra locati

$$\delta^1 A$$

214   et *tam mellitam* saltem gustare loquelam
321   *qui sancti* tumulo servaret membra sacrata
333   illic *hi soli* quia damnantur capitali
397   manda *pro signo* saltim laedi cute summa

$$\zeta A$$

103   talia *pestifero latrando* verbula rostro
293   hic magis *offensus, penitus* fuerat quia victus
357   ardescens *sancti mercari* corpus amandi

$$\zeta B$$

233   tandem *Pelagium nimium* mandavit amandum

HUGO OF LIÈGE

Peregrinarius 1–2000

	A		B	
α^1	17	0.85	3	0.15
α^2	3	0.15	1	0.05
α^3	5	0.25	–	
α^4	2	0.10	3	0.15
α^5	1	0.05	–	
β^1	5	0.25	–	
β^2	5	0.25	–	
γ^1	5	0.25	1	0.05
γ^2	16	0.80	2	0.10
γ^3	–		3	0.15
δ^1	15	0.75	2	0.10
ζ	10	0.50	2	0.18

α¹ A

187	ecce vides *quanto Philippo* Francia rege
188	*regina quanta* styrpeque dives ovet
345	illa tamen *decori regali* singula cedunt
464	sub *iusto titulo,* longa fideque bona
532	*deletas tabulas* summus araret item
680	*emissus corvus* non rediturus abit
694	*intincta virga* dulcia mellis edens
751	quorum *mordaces dentes* sunt arma, sagitte
783	interdum *motus violentus,* quod grave sursum
807	*paccatis nobis* Pluto superabitur hostis
834	proque *bonis habitis* senciet usque malum
970	*pressuras primas* angariasque timens
1088	*salvandas animas* perdere Pluto dolet
1248	in *proprios humeros* impositurus ovem
1482	verba *viris tantis* vociferanda notant
1784	*summi iudicii* venerit illa dies
1867	*ergo pro tanto vestro* concive futuro

α¹ B

788	non *Romam magnam* condidit una dies
1565	scilicet *officium divinum* prepedientes
1923	*dum pomum vetitum* colubro fallente momordit

α² A

741	unitis *uni matri* se fratribus unit
836	que *nulli simili* convaluere, preces
1809	*bis senis signis* in eo septemque planetis

α² B

484	per *totum mundum* iurgia cuncta movent

α³ A

144	*laxatis loris* cedere tempto – moror
219	quomodo *sedatis funestis* denique guerris
1486	*finitis verbis,* ecce repente duos
1707	ut procul *amotis viciis* virtutibus actis
1713	*accepta venia* reges cum Pace recedunt

α⁴ A

911	sopite guerre fuerint et pax sit ubique
1803	illius *rami pulcri,* nimius quoque fructus

α⁴ B

259	*alterutrum sanctum* si dicam, 'devio' dicent
262	*defunctum sanctum* dicere iura vetant
300	*nullum suspectum* iura bonique ferunt

α⁵ A

772	in *vasis captis* robore vina bibens

β¹ A

51	accedensque *meo pretacto* corde iacentem
124	atque *tuis humeris* fers titubantis onus
595	quodque *sui nati* non suggunt ubera matris
1778	proque *suo sponso* sponsa precetur idem
1935	totus namque *tuus thezaurus* traditus illi

β² A

538	qui *nostras guerras* bellaque nostra fovent
613	primitus ut *nostris sedatis* litibus unum
1569	illos *iusticie vestre* pellendo flagellis
1609	quomodo *pacifici vestri* rexere, videte
1867	*ergo pro tanto vestro* concive futuro

γ¹ A

315	gratibus *exhibitis, loris* sinibusque remissis
539	fronte *columbina, volpina* mente gerentes
949	aere garrit *avis, piscis* lascivit in unda
1606	rex *qualis, talis* subditus eius erit
1885	officio *parili simili* tibi nomine presis

γ¹ B

923	quaque sacer *generum, socerum* gener atque nepotem

γ² A

29	spes *desolatis, mestis* solamen, egenti
215	nobilis *insignis fortis* constans preciosa
217	*vinetis granis silvis pratis* fluviisque
352	quodque *duci regi* pontificique placet
851	celum terra *crepent, hylarescent* pontus et aer
867	rusticus *urbanus, populus* plebs, quelibet aetas
1041	Eufrates *Ligeris Iordanis* Secana Nilus
1053	prefrondet *platanus buxus* sambucus oliva
1057	parva mirica frutex *cinus dumus* sicamorus

1058   nec *rampnus dumus* ruscus, ut ante, nocent
1064   sicque *rosas violas* lilia terra capit
1072   *Garganus Pyndus,* sic Therebyntus Eryx
1104   *laudat glorificat* conveneratur amat
1124   plebs *populus clerus* religioque videns
1181   hac mutus *loquitur, graditur* claudus prius audit
1535   *oppressis viduis pupillis* et peregrinis

$$\gamma^2\,\text{B}$$

1673   *Pacem Clementem* regesque dehinc venerantes
1884   post *Linum Cletum* quarte, secunde Petrum

$$\gamma^3\,\text{B}$$

12   *rerum principium* finis, origo boni
376   *servorum servum* te, pater alme, voco
732   *iustorum scutum,* magnus Adulphus adest

$$\delta^1\,\text{A}$$

142   *quas Musas* et quem nos Elycona vocat
184   *a terra* nubes ut fuget atque mari
419   quid prodest *homini, si mundi* cuncta lucretur
536   *pro libito* saperet manna cibusque poli
608   *quo preconcesso* bella parare licet
988   *quo rupto* bigamus nunc magis esse cavet
1142   *pro nostro* dabitur vera labore quies
1413   subticet *hiis dictis* paulisper loraque captans
1437   protinus *hiis dictis* modicum regina propinquans
1445   *quo facto* regi Pax ramos prebet olive
1582   cetera *pro nichilo* computo vestra bona
1603   've terre', dicit Salomon, 'cuius puer est rex
1808   ortum *que specie, que* bonitate beat
1867   *ergo pro tanto vestro* concive futuro
1887   *o sponso tali Felici* foedere iuncta

$$\delta^1\,\text{B}$$

1923   *dum pomum vetitum* colubro fallente momordit
1968   que pridem, *primum dum* medicina fuit

$$\zeta\,\text{A}$$

17   o regina *poli, peccati* nescia, cuius
704   *narcisus Davus* fit, saliunca rosa
710   ecce *mee munde* sunt velut ille manus

902	*psalterii simili* grata canore favo
1280	*viderimus, minimus* corde quibusque patet
1423	a *sexto venio* papa Clemente, libelli
1799	*aspectus cuius* medio protenditur orbis
1887	*o sponso tali Felici* foedere iuncta
1902	grata *tibi, matri* concilianda sue
1923	*dum pomum vetitum* colubro fallente momordit

ζB

| 230 | *tantorum, quantum* singula gesta petunt |
| 421 | heu quis, ut in *messem falcem* ponas alienam |

MATTHEW OF VENDÔME

	A		B	
		Epistule (1808 lines)		
α^1	4	0.22	5	0.27
α^3	2	0.11	–	
α^4	1	0.06	–	
α^5	1	0.06	–	
β^1	1	0.06	1	0.06
β^2	–		1	0.06
γ^1	3	0.17	–	
γ^2	8	0.44	3	0.17
γ^3	–		1	0.06
δ^1	17	0.94	3	0.17
ζ	15	0.83	4	0.22
		Tobias (2226 lines)		
α^1	10	0.45	5	0.22
α^2	3	0.13	–	
α^3	2	0.09	–	
α^4	1	0.04	1	0.04
β^2	2	0.09	1	0.04
γ^1	1	0.04	1	0.04
γ^2	11	0.50	2	0.09
γ^3	1	0.04	–	
δ^1	19	0.89	8	0.36
δ^2	1	0.04	–	
ζ	13	0.58	4	0.18

The relatively modest alpha counts are attributable to Matthew's staccato
style of composition, producing an abundance of couplets like *Epist.*
2.2.131 f. psalmorum series, crepidarum murmura, pellis/immemor etatis, pannicu-
losa, rigens, which do not lend themselves to noun-plus-attribute combinations.

Epistule

α^1 A

2.2.4	*voti legitimi* sollicitare statum
2.7.71	ad *monachos albos* volui transire, sed absit
2.9.7	gaudeo quod *teneros annos* canescere sensu
2.12.32	*communis generis* significata gerat

α^1 B

1.2.65	*fermentum modicum* massam corrumpit, in omnes
1.3.19	in *vetitum cursum* protendo, nitor arenas
1.7.19	*humanum precium* precedis: celibe vita
2.1.111	*activum verbum* non est in amantibus, immo
2.1.129	est amor *egregium verbum,* constructio cuius

α^3 A

2.4.11	*sopito stimulo* caude promissa tepescunt
2.8.41	*excepto modulo* victus innata scolari

α^4 A

2.6.17	reddit *eos tepidos* vobis elatio vene

α^5 A

2.5.30	*litterulis pressis* sollicitudo datur

β^1 A

1.6.21	ex qua peste *meus questus* suppullulat, audi

β^1 B

2.9.70	iste *tuum precium* depreciare queat

β^2 B

1.2.35	vester apex *vestrum vicium* declarat, in usum

$$\gamma^1\,A$$

1.1.88	aula *dei, cleri* sanctificata domus
1.2.20	*compatior, videor* vulnera vestra pati
2.1.53	pascitur *intuitus, tactus* ieiunat: amicat

$$\gamma^2\,A$$

1.prol.9	qua duce mens *loquitur, auditur* qui tacet, absens
1.1.53	te pastore *rati, tuti* te praeside, freti
2.3.7	compatiens *misero, mesto* iocunda, dolenti
2.4.98	*ditat, letificat* vultus, amicat odor
2.9.38	*concelebrant, reprobant* dogmata, probra colunt
2.9.42	*blandiciis, laqueis* implicat, ungue rapit
2.10.19	lactea *blandiciis, verbis* mellita, tenore
2.12.7	obsequio *blandus, discretus* mente, fidelis

$$\gamma^2\,B$$

1.7.32	*igniculum, radium* de radiante feram
2.9.12	*augmentum, precium* ditat, honorat honor
2.13.8	*eloquium studium* dogmata fama decus

$$\gamma^3\,B$$

1.1.16	*harum litigium* pacificante modo

$$\delta^1\,A$$

1.1.12	linea *iusticie, te* mediante regens
1.2.38	*qui stomachi* gaudent uberiore globo
1.2.75	*pro Christo* gaudete pati vobisque talenta
1.2.77	vobis *pro numero* reddatur talio: quisquis
1.3.73	dictandi *species tres* disciplina secundum
1.5.57	*he rotule* cupidum subvertunt: spes timor ira
1.7.13	ut *se iustitie* pietas votiva maritet
2.1.123	vix impendo fidem *liquide, que* nata liquari
2.3.25	ve mihi, *ve misere, que* sollicitare pudicam
2.4.1	digna *laboranti si* reddat premia, facto
2.4.22	non placet: *a nulla* plaustra reflectis humo
2.4.53	allicit ut redimat, *cupidi si* dicta tepescant
2.6.13	*pro modulo* propera: ieiunus prandia debes
2.7.55	*hi personali* precedunt gutture, ventris
2.8.38	*nummos quos* dederas Parisiense solum
2.8.61	si memor es memoris, *patrii si* vernat amoris
2.9.76	ut pater *a patria* proprietate cadam

δ¹ B

2.2.94	malo *pudiciciam quam* tua vota sequi
2.4.25	miror te *stolidum, dum* relliquias popularis
2.5.18	exulis *heredem spem* iubet esse metus

ζ A

1.1.1	summo *pontifici cleri* conventus, odorem
1.1.109	in culmen *domini, famuli* pressura redundat
2.prol.5	*scribendo redimo* dispendia temporis, usum
2.1.31	regna reor *Veneris pubis* ieiuna, saporis
2.1.37	scemata *dispono, falero* vel sicut Ulixes
2.1.109	heu patior nec *ago: modico* ieiunat ab actu
2.1.133	quippe *nisi voti* socer intercessor habundet
2.2.114	*excludo; placeo,* displicet, uro, furit
2.3.62	unde *mihi potui* conciliare dapes
2.6.69	spem refove *tutus: precibus* cum venerit hora
2.7.9	sacri *delicias dispensas* pectoris, actu
2.8.37	*indigeo: careo* libris et vestibus, hausit
2.11.37	excitor, *exsurgo, timeo* tibi, consule fame
2.12.24	*obsequio cupio* posse bonumque bono
2.13.5	cuius honor, *cuius titulus* presignis ab hoste

ζ B

1.1.9	gemma *sacerdotum, speculum* telluris, honoris
2.prol.3	vas *figulum, scriptum* scriptorem, fabrica fabrum
2.4.46	*liliolum, precium* sordet, acescit odor
2.4.84	*coniugium vinclum* conubiale fugit

Tobias

α¹ A

113	*primevis tunicis* rosa dedignata teneri
135	sole *novus radius* prodit, lux lumine, dignus
259	*iusto proposito* Tobias constat, in actus
299	Chamo vel *rigido freno* constringe rebelles
744	*arbitrio proprio* proprietate caret
1103	extrahit, *angelicis monitis* exenterat, inde
1148	*multiplicis prolis* fructificare queas
1217	virgo timet, *teneris lacrimis* exuberat: usum
1977	heu *quantis viciis,* quanta gravitate moderni
2110	*pentametris elegis* Vindocinensis amat

$$\alpha^1\,B$$

52	ut *patrum veterum* testificatur apex
558	ex *patrum veterum* traditione potes
968	*maternum tumulum* continuare stude
1243	ultra *femineum sexum* virguncula vernat
1389	tanta fides, *tantum precium* fastidit honorem

$$\alpha^2\,A$$

1196	debet *ei soli* virginitatis opes
1335	inde *duas vitulas* pingues oviumque maritos
1406	bisque *duos servos* eligit, haurit iter

$$\alpha^3\,A$$

| 1213 | hinc *factis tenebris* sopito sole, iubente |
| 2043 | *defunctis soceris* processu temporis, ambos |

$$\alpha^4\,A$$

| 41 | vivite *felices, fratres,* quos corpore solo |

$$\alpha^4\,B$$

| 1295 | huic *Evam sociam* tua dispensatio iungi |

$$\beta^2\,A$$

| 1259 | *ecclesie nostre* quantum synagoga fidelis |
| 1793 | quippe *pater noster,* dum se sperni videt, hosti |

$$\beta^2\,B$$

| 1207 | *vestrum coniugium* confirmet, prole beatos |

$$\gamma^1\,A$$

| 1357 | prestant nulla *gule, nature* paucula; vite |

$$\gamma^1\,B$$

| 256 | *obsequium, meritum* garrulitate premit |

$$\gamma^2\,A$$

54	*metrificat, reprobat* livor, amicus habet
89	odit *amat reprobat* probat execratur adorat
109	condolet *afflictis, miseris* blanditur, egenis
197	*iustos innocuos functos* premit enecat arcet
837	visus *odoratus auditus* tactus amicus
935	sperne *malignantes, sapientes* consule, pravos
1231	nubit casta *pio, iusto* devota puella

1804	*deliciis, viciis* ebria, legis inops
1889	increpat *elatos, iratos* mitigat, egris
2079	visne *probis iustis* vel sanctis parcere? parce
2080	parce: *probus iustus* sanctus uterque iacet

γ^2 B

34	*portum, naufragium* pelle, medere rati
2063	eclipsis *radium, speculum* cinis, umbra nitorem

γ^3 A

1781	*electi Domini,* Dominum benedicite, corde

δ^1 A

83	ad *vitulos, quos* Iheroboam rex fecerat, ire
575	effuge *tres hostes* anime, versutia quorum
577	*tres hostes* hominis − demon caro mundus − acerbant
779	sic des *pro modulo* census, ne prodiga dandi
857	*hos famulos* cogas tibi deservire: ministri
871	ne credas titulis *fame, te* consule, crede
981	*his monitis* accede patris: te credere patris
987	*pro modulo,* pater, etatis parere paratus
1266	*prolis? quis Veneris* armiger esse negat
1461	ne timeas: vir *qui pueri* moderatur harenas
1519	*his dictis* iteratur 'ave', dant oscula fletu
1572	tentativa *vie que* solet esse, manus
1635	*his dictis* genitor, natus, Raphaele vocato
1699	*his dictis* fugit ex oculis nec luminis usus
1912	servili sed *eo quo* reveretur, amat
2047	*his prelibatis* Tobias vendicat heres
2127	*hos elegos* distingue: patent; distinctio cesset
2146	mater, *ubi patrui* sed patris ossa iacent
2184	et *plures; tres* sunt, non tria, tres et idem

δ^1 B

369	non intres in *iudicium cum* paupere dives
371	non intres in *iudicium cum* pulvere fortis
448	*quam famam* vicio sordidiore premi
767	venditur *arbitrium, dum* vivitur ex alieno
1006	esse, sue *comitem quem* rogat esse vie
1492	precipitare *moram quam* parricida mori
1677	de numero septem *sum spirituum,* quibus esse
2197	felix *coniugium: dum* se sacra verba maritant

δ^2 A

1187   cursitat *huc illuc* Raguel, discumbere letus

ζ A

47   aspirate *stilo; vestro* mendica phaselus
106   *primitias, decimas* compatiente manu
403   *Asmo demonio* nomen – pars ultima vocis
543   hoc precor, hoc *iubeo: toto* conamine, tota
566   *propositi, facti* previa, mentis iter
809   da laudes *Domino iocundo* tempore, lauda
976   mente *Deus, cumulus* prosperitatis adest
1210   *coniugii scripti* testificante nota
1459   Anna tace, *noli turbari:* sospite nato
1498   *ancillas, reliquas* que reticentur opes
1835   consolare, *queri noli,* letare: doloris
1896   languet vita *senis, mortis* cognata senectus
2201   intus et *exterius totus,* quia virgine totus

ζ B

694   *exilium, celum* patria, vita Deus
1158   *hospicium, factum* purpurat oris honor
1386   *consilium, morum* gratia, firma fides
1716   vox *sensum, nucleum* testa, favilla iubar

NIGEL DE LONGCHAMPS

Miracula Virginis Mariae (2690 lines)

	A		B	
α^1	9	0.33	4	0.15
α^2	1	0.04	–	
α^3	1	0.04	–	
α^4	–		2	0.07
α^5	–		1	0.04
α^7	2	0.07	–	
β^1	8	0.30	2	0.07
γ^2	1	0.04	2	0.07
γ^3	–		1	0.04
δ^1	14	0.52	6	0.22
ε	2	0.07	–	
ζ	17	0.64	9	0.33

$$\alpha^1$$

145	demon ad hec: *homini tali* Chistumque colenti
347	*delitiis variis* opibusque fluens preciosis
470	*magnus Basilius* obvius esset ei
600	que *meritis dignis* munera digna dedit
724	curre *vias longas* per brevitatis iter
821	triste sed et *tristes fratres* sallendo frequentant
1667	salmorum *numerus varius* varias notat herbas
1710	*verbis virgineis* facta dedere fidem
2107	protinus ergo *sacri babtismi* fonte renatus

$$\alpha^1\,B$$

1127	*uxorem sterilem* causatur habere maritus
1901	ergo *virum iustum*, nullo prohibente tuorum
2011	ergo *techam modicam* subtili vimine textam
2461	et quia *femineum sexum* corpusque tenellum

$$\alpha^2\,A$$

1660	*et sexti salmi* signa sacrata gerit

$$\alpha^3\,A$$

1952	*absumptis preciis*, nil valuere preces

$$\alpha^4\,B$$

1557	cumque puer *puerum vivum* verumque putaret
2654	*inventum puerum* nuntiat atque senem

$$\alpha^5\,B$$

2569	dixit et *exceptum puerum* de ventre parentis

$$\alpha^7\,A$$

9	que velut in *celis superis* comitata choreis
1251	cumque satis *regi toti* placuisset et urbi

$$\beta^1\,A$$

793	cumque *suis viciis* miser insenuisset et annis
992	ille *meus monachus* cras adeundus erit
1053	ceperat ergo *suo tanto* de crimine fidus
1081	rex, in monte *tuo sancto* quisnam requiescet
1573	cui puer: 'iste *meus socius*' digitoque sacriste
1755	dampna *sue lingue* multa virtute redemit
1879	ille *meus famulus* et cancellarius idem
2587	sit moderata *tui discreti* norma rigoris

β¹ B

277	nonne *meum natum* blasfemus et ipse negasti
1075	cumque fuisse *suum monachum* Petrus hunc meminisset

γ² A

1167	alloquium *tactus visus* locus aptus amori

γ² B

595	vestis *candorem speciem* decus atque decorum
1419	vas Domino gratum, vas *mundum sanctificatum*

γ³ B

179	luxus *opum numerum,* mensuram copia rerum

δ¹ A

169	solvitur *his gestis,* tanto gavisa triumpho
174	et, velut ante *fuit, fit* vicedomnus item
513	defluit in *lacrimas quas* aut timor aut dolor urget
1227	consulit ergo *suos quos* possidet ille nocendi
1564	vim faciens *puero quo* valet ille modo
1607	quid, *quo consilio,* quare vel qua ratione
1697	absit ab *his labiis* fetor furor et dolor omnis
1853	sic voluit *cogi qui* sic est forte coactus
1903	quod si distuleris, *pro certo* te moriturum
2160	omnia *si mundi* tu michi regna dares
2256	causantur *subitas has* in amore moras
2353	transiit *his dictis* tristemque reliquid amantem
2460	amplius *his aliis* sollicitare studet
2517	ergo recepta *loco quo* multiplicare Marie

δ¹ B

367	cuius ad *exemplum dum* se componit amando
1001	noveris ergo *piam quam* conspicis esse Mariam
1033	*vitam quam* fuerat monachi sub veste professus
1341	currit in *incertum, dum* spe frustratur inani
2154	expedit et *Christum cum* genetrice sua
2563	sed quia *sum memorum* memor indefessa meorum

ε A

1096	*pro quo* decrevit fundere virgo preces
1641	hunc igitur, sompno dum *se de* nocte dedisset

ζ A

95	ira *licet simulet* modico sub tempore vultum
150	*confisi Christi* de pietate sui

355	testis adest genti pariterque *Dei genetrici*
969	sub specie *triplici tauri* canis atque leonis
1005	dixit et hoc *dicto monacho* mundoque relicto
1089	obvia fit *matri venienti* gloria nati
1107	grex congaudet ovi de fauce *lupi redeunti*
1145	nascitur ergo *patri sterili* de coniuge natus
1185	unicus est *matri defuncti* forma mariti
1283	dixerat. at *cuncti communi* voce resistunt
1358	*concepi; peperi* crimen onusque mihi
1361	addo scelus *sceleri, veteri* nova dampna furori
1489	simplicitas *pueri mentiri* nescia veri
1589	hoc speciale *decus primus* dedit iste Marie
1747	cuius amor *tantus, cuius* miseratio tanta
1893	de pastore *lupus, factus* de presule predo
2469	uritur igne *novo subito* succensa calore

ζB

55	contulerant in *eum cunctorum* vota favorem
361	huius in *obsequium totum* se devovet: huius
635	ad deitatis *opem tandem* sibi confugiendum
731	fugerat in *claustrum, mundum* fugiendo sequentem
1174	in *thalamum puerum* suscipit inque thorum
1487	contigit ergo *domum puerum* de more reversum
2515	hanc et non *aliam solam* superesse Mariam
2534	*offensum placidum* redde, benigna, Deum
2633	confessusque *palam quoniam* peccasset in illam

PSEUDO-OVIDIUS

De vetula I (818 lines)

	A		B	
α^1	17	2.08	2	0.24
α^2	6	0.73	–	
α^3	3	0.36	–	
α^4	–		1	0.12
α^5	2	0.24	–	
γ^2	1	0.12	–	
γ^3	1	0.12	1	0.12
δ^1	12	1.45	7	0.86
ε	3	0.36	–	
ζ	5	0.61	3	0.36

α¹ A

2	*femineus sexus,* sine quo nec vivere posse
33	et iuvenescebam, quotiens *mento reparato*
107	hic *viduus ramus* foliis facientibus umbram
264	*ventilabro moto,* possim stabilone ligato
267	donec in *alatas caligas* et pyramidales
272	sicut iectigat is, cuius *nervi resoluti*
304	donec in *insidias pretensas* retiolorum
362	cum *solus casus* in talibus inveniatur
411	*compositi numeri* nascuntur, non tamen eque
446	*predicti numeri; si* vero sit unus eorum
456	*compositos numeros,* quibus est lusoribus usus
625	et *medius cursus* est idem semper eorum
636	sunt *alii ludi parvi,* quos scire puellas
667	seu scaci portant, et sunt *acies bicolores*
674	*tetragoni reliqui,* nisi quod duo sunt ibi reges
686	et sunt ex *numeris quadratis* omnibus ambe
687	quod potes ex *tabula subiecta* noscere plane

α¹ B

50	naturaliter est ad *sexum femineum,* qui
662	toti *particulam dictam* patris a quotitate

α² A

241	*delicias istas* et gaudia tanta docere
374	*divitias alias* et agros agris cumulasse
479	adde quod in *multis ludis* quicumque lucratur
577	est *alius ludus* scacorum ludus, Ulixis
636	sunt *alii ludi parvi,* quos scire puellas
804	*istos indignos* indignus promovet iste

α³ A

166	*aversis oculis* teneri puerilia nati
389	*vini potandi* maturos duxit ad annos
488	nam *posito fato* libertas arbitrii non

α⁴ B

490	est *fatum fatuum; fatui qui* fata sequuntur

α⁵ A

594	et tamen est *numerus finitus* motibus ipsis
596	sic *ludus factus* motus caelestis ad instar

$$\gamma^2\,A$$

204   sed *potuit scivit voluit* tantum indere, nosque

$$\gamma^3\,A$$

425   hi sunt sex et quinquaginta *modi veniendi*

$$\gamma^3\,B$$

708   discerni per *iudicium phisiagnomicorum*

$$\delta^1\,A$$

38   non pro se, sed *pro signato:* silva viriles
339   retibus *hos; illos* hamis illosque sagena
385   *pro plumbo* lateres mutabat, ferrea vendi
396   sed qui *pro ludo* sua dispergit, nichil unde
427   nam quando similes fuerint sibi tres *numeri, qui*
446   *predicti numeri; si* vero sit unus eorum
487   si fatum ponas, *fatui qui* fata sequuntur
490   est *fatum fatuum; fatui qui* fata sequuntur
542   *pro modico* liceat vel fortuitu sine dampno
580   vel *belli, si qui* pro vulneribus remanerent
752   qui pacem *pro dimidio* sumptus habuisses
770   *qui populi* vice preficiunt concorditer unum

$$\delta^1\,B$$

17   vellem *quam nullam* quod concordem magis unam
21   *perfectam, quam* si citius fortuna dedisset
91   plusque ferebar ea *quam quoquam* materiali
173   mortem *quam vitam* vix natus tempora complens
529   si *quam captivam* regredi contingeret illuc
747   *quam causam.* non curat enim, quantum tibi constet
809   artes, *nam quoniam* sublimant deteriores

$$\varepsilon\,A$$

16   non placet, hoc renuit; *de me* scio, quod magis unam
154   armaque iusta movet, *qui, si* prepossit, amanti
580   vel *belli, si qui* pro vulneribus remanerent

$$\zeta\,A$$

369   *ludi lucrandi* solius odore moveri
449   sed si dissimiles sunt *omnes, invenies* sex
502   fine *tenus ludus.* nec sola sorte sed arte

716   tunc matheses *sole doctrine* nomine digne
762   alkimiam, *cuius fructus* ditatio tanta

ζ B

79   *inclusum mecum* satis et mirabiliorem
210   *rursum: preceptum* si negligo, non quia iustum
601   in *campum primum* de sex istis saliunt tres

THEODULF

Carmina 1–25 (1716 lines)

	A		B	
α^1	14	0.82	4	0.23
α^2	1	0.06	–	
α^7	1	0.06	–	
β^1	4	0.23	–	
β^2	1	0.06	–	
γ^1	14	0.82	–	
γ^2	3	0.17	–	
γ^3	–		1	0.06
δ^1	14	0.82	2	0.12
δ^2	1	0.06	–	
ε	1	0.06	–	
ζ	3	0.17	1	0.06

α^1 A

1.57   atque *lacessitis sociis* hinc inde ministret
1.189   proque *polo sudo* tetrum est mercatus Avernum
1.256   in *tepidi libri* sistito fine gradum
3.9   proque *bonis propriis* clemens sua praemia donat
7.59   cumque *bonis propriis* veniat fortissimus Astur
10.14   tantum *eius sensus* post bona verba valet
15.13   quo *dictis variis* aptantur dicta decenter
17.69   *divitiae verae* sunt gloria, norma fidei
17.81   atque in *apostolicae vitae* fundamine stare
18.22   *vatidicus lituus,* rex pie Christe, tuus
21.45   *pontifices celebres* et vatum summus Esaias
21.90   cum *doctis sociis* discipulisque suis
25.135   et pia de *sanctis scripturis* dogmata promat
25.243   qui te *mundani regni* rex extulit arce

α¹ B

6.11   *fallacem testem* mendacia saeva canentem
19.15   liberat *oblatam moecham* fallacibus illis
21.103   eruet aut *animam propriam* quis manibus imis
25.106   *lacteolum strophium* haec vehit, illa rubrum

α² A

1.169   quae quamquam *vitiis cunctis* alimenta ministret

α⁷ A

25.201   his bene *patratis, mensis* dapibusque remotis

β¹ A

1.20   deque *sui nigri* flumine lactis alit
1.138   atque *tuis armis* tuta caterva sumus
11.38   proque *suis meritis* ignea flagra pati
23.15   nempe *tuis monitis* extendis viscera terrae

β² A

23.42   *versiculos nostros* sollers reprehendere noli

γ¹ A

1.195   linquitur et *toties, quoties* obsistitur illi
2.229   adsertor *veri, perversi* dogmatis hostis
10.15   carmina plura *queunt, nequeunt* tamen omnia, quamvis
10.37   hic Iuda *peior, melior* te, Petre, videri
12.11   num pro ope *quis panis, lapidis* tribuendo rigorem
12.35   cum nunc aequa *meent, remeent* et iniqua dolenter
13.5   nam mala saepe *bonos, reprobos* bona saepe sequuntur
13.26   ut quid *agat habeat* ultima magna dies
14.17   cultor *agris, castris* et miles, navita ponto
19.17   nam pius hanc *salvat, damnat* peccamina iustus
19.35   stare illum *Stephanus, Marcus* residere profatur
19.51   quosque prius *sanat, satiat* post nectare magno
25.235   rex sua fulcra *petat, habeat* sua mansio quemque
25.239   qui ne quem *offendat, placeat* dilectio Christi

γ² A

1.123   virtutum *genetrix nutrix* defensio sollers
7.29   et quod habet *Moesus Dacus* Thrax Atticus Arcas
25.45   ut veniunt *Abares Arabes* Nomadesque venite

γ^3 B

7.3.11   inque *gregum modulum* procedit parvula turba

δ^1 A

1.260   captivum, *his armis* subdideratque sibi
1.291   saepe etiam per *res similes* conferre medelam
2.61   improbat indoctos *his verbis* sanctus Esaias
2.82   cerea lux *aliis, fis* cinis ipse tibi
2.269   sit tibi secretum *pro pulchro* pondere dogma
12.11   nam pro ope *quis panis lapidis* tribuendo rigorem
16.19   transitus *a laeva* si in dextram ex ordine fiat
17.38   plus cerni *hi sancti* quam tamen esse volunt
17.64   hoc signant *domini qui* residendo volunt
17.73   dum de illis cano *qui sancti* specie tenus extant
17.110   actibus in *saecli qui* sine labe fuit
19.10   quid nescire *dei ni* reprobare velit
19.45   cernit eundo *duos, quos caecos* stando resanat
25.216   litigiosa *lues, res* fera, grande nefas

δ^1 B

2.121   non ego *pontificum, dum* vitam scribo malorum
25.51   ver venit ecce *novum, cum* quo felicia cuncta

δ^2 A

25.155   Nardulus *huc illuc* discurrat perpete gressu

ε A

17.15   sic *tunc: nunc* alio discurrunt omnia cursu

ζ A

2.33   quique *pio cupio, soleo,* vel debeo voto
12.23   nam pia dona *spei tereti* signantur in ovo
14.3   nam loquitur *rebus, taceamus* dogmata quamvis

ζ B

1.255   quae per prata *parum vitiorum,* Musa, measti[1]

---

[1] Repeated v. 313.

MARCUS VALERIUS

Bucolica (451 lines)

	A		B	
α^1	1	0.22	–	
α^7	2	0.44	–	
γ^2	1	0.22	–	
δ^1	2	0.44	1	0.22
ε	–		1	0.22
ζ	4	0.89	1	0.22

α^1 A

2.64 totque *deos medios* totque oscula cara relinquor

α^7 A

3.103 optat aper *silvas, maturas* vinitor uvas
3.105 gaudet apes *calathis, sociatis* vitibus ulmus

γ^2 A

3.68 silve prata *greges segetes* vineta fluenta

δ^1 A

1.6   et pudet *has saturas* non semper cernere fetas
3.52 scis *catulos, quos* sepe petis sub ventre Licisse

δ^1 B

3.40 et certare paras? *timeam quam*, cerne: Licotas

ε B

1.71 aut michi sint fures, fateor: *nam iam* (male sensi)

ζ A

Prol. 1 parva quidem *arbitrio committo* carmina magno
2.26 Eufili, tu *roseas poteras* superare Napeas
2.28 ni *sis equoreis* facta inconstantior undis
3.89 me donec *videat, iactat* convitia divis

ζ B

2.13 vel moriturus *eram. quidnam?* num munere certem

WALTHARII POESIS (1456 lines)

	A		B	
α^1	6	0.42	–	
α^2	1	0.07	–	
α^3	2	0.14	–	
α^4	–		1	0.07
α^7	1	0.07	–	
β^2	1	0.07	–	
γ^1	1	0.07	–	
γ^2	–		1	0.07
γ^3	–		1	0.07
δ^1	13	0.90	16	1.11
ε	7	0.48	–	
ζ	6	0.42	5	0.35

α^1 A

524	contra aquilonenses sive *australes regiones*
537	ipse *oculos tersos* somni glaucomate purgans
546	ut quae non merui *pacto thalamo* sociari
867	*caeligenas animas* Erebi fornace retrudunt
1156	undique *praecisis spinis* simul et paliuris
1376	belliger ut *frameae murcatae* fragmina vidit

α^2 A

| 64 | ibant legati *totis gladiis* spoliati |

α^3 A

| 202 | et *versis scutis* laxisque fruuntur habenis |
| 244 | nullus adest *nobis exceptis* namque duobus |

α^4 B

| 760 | hic ubi *Waltharium promptum* videt esse duello |

α^7 A

| 1384 | gentibus ac *populis multis* suspecta tyrannis |

β^2 A

| 376 | en hodie *imperii vestri* cecidisse columnam |

γ^1 A

205   deicit *obstantes, fugientes* proterit, usque

γ^2 B

1273   unice enim *carum, rutilum, blandum, pretiosum*

γ^3 B

266   his *armillarum tantum* da Pannonicarum

δ^1 A

257   vestrum velle meum, *solis his* aestuo rebus
372   o *detestandas, quas* heri sumpsimus, escas
434   illic *pro naulo* pisces dedit antea captos
505   Hiltgunt, et nebulam *si tolli* videris atram
594   *his auscultatis* suggesserat hoc adulescens
656   aut mihi *pro lucro* quicquam donaverat ille
749   accurrit iuvenis et *ei vi* diripit ensem
979   en *pro calvitio* capitis te vertice fraudo
1006   *quo dempto* vivus facile caperetur ab ipsis
1123   tum pugnare putes, *belli si*, rex, tibi mens est
1276   idcirco gazam *cupio pro* foedere nullam
1411   est athleta bonus, *fidei si* iura reservet
1443   *his dictis* pactum renovant iterato coactum

δ^1 B

25   hoc melius fore, *quam vitam* simul ac regionem
78   *quem subolem* sexus narrant habuisse virilis
160   me vinclo permitte *meam iam* ducere vitam
218   si bene res vergant, *tum demum* forte requirunt
471   *gazam, quam* Gibicho regi transmisit eoo
490   venerat in saltum iam *tum Vosagum vocitatum*
503   bellica *tum demum* deponens pondera, dixit
627   me petit atque *oculum cum* dentibus eruit unum
696   qui *dum Waltharium* multo terrore videret
700   non ego *iam gazam* nec rerum quidque tuarum
820   reddes: *tum demum scelerum* cruciamina pendes
941   *tum primum* Franci ceperunt forte morari
948   ante mori sum, *Wormatiam quam* talibus actis
1135   *tum secum* sapiens coepit tractare satelles
1193   armillas *tantum, cum* bullis baltea et enses
1317   et *iam comprensam* sensim subtraxerat illam

ε A

229	ambo etenim norant *de se* sponsalia facta
233	inter se nostra *de re* fecere futura
447	et, cum venisset, *de re* quaesitus eadem
552	qui *me de* variis eduxit saepe periclis
980	ne fiat ista tuae *de me* iactantia sponsae
1042	fors tibi victoriam *de me,* non inclita virtus
1244	sperabam, fateor, *de te,* sed denique fallor

ζ A

14	quorum rex *Gibicho solio* pollebat in alto
151	vinciar *inprimis curis* et amore puellae
508	ne excutias *somno subito,* mi cara, caveto
762	sive per *aerias fallas,* maledicte, figuras
874	cui nec rapta *spei pueri* ludicra dedisti
1331	sic dum *Waltherius vulnus* cavet, ille resurgit

ζ B

99	virginis et *curam reginam* mandat habere
392	decidit in *lectum, verum* nec lumina clausit
490	venerat in saltum iam *tum Vosagum vocitatum*
647	regi *Francorum totum* transmitte metallum
820	reddes: *tum demum scelerum* cruciamina pendes

YSINGRIMUS I–IV (3994 lines)

	A		B	
α^1	7	0.18	3	0.08
α^2	2	0.05	–	
α^3	2	0.05	–	
α^4	–		1	0.03
α^5	1	0.03	–	
α^7	1	0.03	–	
β^2	1	0.03	–	
γ^1	5	0.13	2	0.05
γ^2	2	0.05	–	
γ^3	–		3	0.08
δ^1	28	0.70	11	0.28
ε	5	0.13	–	
ζ	6	0.15	3	0.08

$$\alpha^1 A$$

1.2	*ieiunis natis* quereret atque sibi
1.742	in *primis feriis*, et 'kyri' vulgus 'ole'
2.99	*oratrus fratrus*, Paz vobas clamat et infert
3.1020	*agnini iuguli* vulnere tutus erat
4.655	a *caris sociis* huc tu conductus es usque
4.809	sic *asinus victus* Reinardi vicerat astu
4.942	*eximiis proavis* inferiora facit

$$\alpha^1 B$$

1.293	lumina trans *humerum dextrum* torquere parabat
2.658	*dilectum patruum* decoriare parat
4.865	neve *viam sacram*, quamvis sit dura, timete

$$\alpha^2 A$$

3.313	tunc *multas soleas* nec hiantes vulnere pauco
4.354	gensne *alii genti* faverit ulla magis

$$\alpha^3 A$$

1.289	*erectis oculis* absentem denique sentit
3.694	*supposito titulo*, sillaba qualis erit

$$\alpha^4 B$$

4.951	egregiam *prolem maiorem* patribus esse

$$\alpha^5 A$$

1.737	iamque *sacerdotis stantis* secus atria gallum

$$\alpha^7 A$$

1.959	his tandem *lacrimis mestis* compassa querelis

$$\beta^2 A$$

2.638	ʼexuviis vestris, spes caret illa fide

$$\gamma^1 A$$

2.105	dumque *levat, clamat*, quem nec mutire decebat
2.223	ut suus est *modicis, nimiis* sic terminus ausis
3.73	*desipiat sapiat, vivat* moriatur egenus
3.87	accipiens *reddat, timeat* quicumque timetur
4.800	qui *ruerant, poterant*, fit timor atque tremor

γ^1 B

1.230	ire *velim nolim,* vis, ierisque feram
4.90	ibo, *queam nequeam,* qualibet arte ferar

γ^2 A

1.561	res *proprias medias alienas* credis easdem
3.424	*impingas, vellas,* denegat illa sequi

γ^3 B

2.186	sive *horum neutrum* fecerit, unus ero
2.301	*amborum numerum* Bernardi in fronte rigere
3.422	*herbarum modicum* conficit hora brevis

δ^1 A

1.87	huc ergo *cupide, ne* sero intrasse queraris
1.119	querat an arte aliqua *redimi, qui* septus in arto est
1.278	sed numquam venturus *eo quo* creditur isse
1.297	prodiit *a leva* rediens, oculosque latentem
1.443	*his dictis* abiit Reinardus, fratre relicto
1.816	vis modo *restitui, si* potes, omen habes
1.821	stent igitur stantes, *strati, si* copia, surgant
1.981	sacra tibi *his sacris* dabimus, que verbera si non
1.1017	qualibet *a summa* sene grex numerante gradatim
2.62	dicitur *a furca* quem rapuisse deus
2.95	hos anus atque *alios, quos* est mora dicere, sanctos
2.146	*pro naso* pape stare tulisset ibi
2.326	alternant *dubie, que* facienda forent
3.387	finiturus *eo quo* rex magis utitur ore
3.447	pelle *lupi, qui* dimidium tribus addidit annis
3.492	consiliis, *regi si* favet ille, valet
3.541	*his positis,* Reinarde, doce quid deinde sequatur
3.606	sed gravis *offense te* tenet aula reum
3.756	*pro nichilo* vidi sic trepidare probum
3.1029	quam *poteris bis* tincta rubet, de qua iste superbit
4.218	nec quis in *his lucis* invenit ova duo
4.252	quid velit *ignorat, dat* quibus ipsa caret
4.477	per *sanctos, quos* quero, nisi mea iussa sequaris
4.752	Mantica, *quo genero* Sualmo superbus erat
4.862	non vos *pro modico* crimine trudo foras
4.873	quin, ne *nos famulos* tibi dedignere, veremur
4.884	*his dictis* celerem corripuere viam
4.921	per *sanctos quos* queris, abi, Reinarde! manebo

δ¹ B

1.400	tundatur *ferrum, dum* novus ignis inest
1.577	hec tria cur *fugiam, quam* congrua causa sit audi
1.603	nec potior *quisquam quam* tu michi crederis esse
1.996	nec minus hoc *cuiquam quam* sibi velle lucrum
2.176	qua latro *calicem, quem* dedit, ipse bibat
2.266	malo quoque ut *queram quam* caruisse feram
3.505	'patrue' clamabat 'numquam tam mira notavi
3.550	*quam faveam* patruo notificabo semel
3.768	certius *antidotum sum* meditatus ego
3.782	si regis *curam, quam* profiteris, habes
4.159	nec michi *tam quosquam quam* vos invisere dulce est

ε A

1.806	qui queritur *de te* perpetiatur idem
3.375	spes michi *res, spes* sola comes, mox curro Salernum
3.476	non dubitat *de se* dicier ista senex
4.949	fama nichil *de te* perhibet, nec scimus iniquo
4.979	esurio, servabo fidem, nil vendico *de te*

ζ A

1.944	*Reinardi credi* forma fuisse potest
2.359	quas, domine, hic *epulas mandas* tibi? vivimus herbis
3.1081	stulte, quid *emendas, ignoras?'* namque rogabat
4.5	*Rearidus cervus,* suspectum ductor in agmen
4.50	aufer *opes, exspes,* redde, petulcus erit
4.317	quis me, inquit, *Sattanas lupicidas* traxit ad istos

ζ B

1.797	pura fides *etiam personam* pauperis ornat
3.1161	si vis ferre *moram, postquam* sudabo, resumptis
4.889	tunc *baculum secum* peramque prehendit abitque

NEO-LATIN

In the neo-Latin poets as here listed alphas are on average about as common
as in the religious ones of late antiquity: in 19671 lines 102 α^1s (0.52%) and
28 α^{2-6}s (0.14%). Omegas, with 242 (1.23%), somewhat less common.[1]

The surprise comes with the announcement that three of those surveyed,
and these the most accomplished – Politian, Sannazarius, Faustus Andrelinus
– stand out from the rest with scores that disclose a quite classical aversion
to alpha homs:

	lines	α^1	α^{2-6}	ω
Politian	2 357	2 (0.08%)	1 (0.04%)	25 (1.06%)
Sannazarius	4 337	8 (0.18%)	1 (0.02%)	53 (1.22%)
Andrelinus	1 311	1 (0.07%)	–	11 (0.84%)
Others	11,666	91 (0.78%)	25 (0.21%)	153 (1.31%)

Without the three puritans we get an alpha count that rivals the medievals.

Again the question: did the three know? Clearly Andrelinus' imitator
Arnolletus did not. It would rather seem that by some kind of aesthetic
osmosis their dedication to classical models developed in them a like sensi-
tivity, of which they like their models were unconscious or almost so.

Andrelinus (1311)	0.07 (0 + 0.07)	0.07 (0 + 0.07)	0.83 (0.53 + 0.30)
Arnolletus (471)	2.31 (1.89 + 0.42)	2.54 (2.12 + 0.42)	2.10 (1.46 + 0.64)
Buchanan			
De sphaera (c. 2000)	0.90 (0.80 + 0.10)	1.15 (0.95 + 0.20)	0.90 (0.50 + 0.40)
Franciscanus (c. 900)	0.89 (0.89 + 0)	1.22 (1.22 + 0)	1.66 (1.33 + 0.33)
Naugerius (784)	0.64 (0.64 + 0)	0.64 (0.64 + 0)	1.16 (0.90 + 0.26)
Petrarca (1953)	0.81 (0.61 + 0.20)	0.93 (0.67 + 0.26)	1.38 (0.92 + 0.46)
Politianus (2357)	0.08 (0.08 + 0)	0.12 (0.08 + 0.04)	1.06 (0.76 + 0.30)
Pontanus (1321)	0.68 (0.68 + 0)	0.68 (0.68 + 0)	0.46 (0.38 + 0.08)
Rakovský (2632)	0.49 (0.45 + 0.04)	0.99 (0.80 + 0.19)	1.55 (1.10 + 0.45)
Sannazarius			
De partu Virginis (1443)	0.14 (0.14 + 0)	0.14 (0.14 + 0)	1.46 (1.32 + 0.14)
Elegies (1506)	0.20 (0.20 + 0)	0.20 (0.20 + 0)	1.06 (0.80 + 0.26)
Others (1388)	0.22 (0.22 + 0)	0.29 (0.29 + 0)	1.17 (1.03 + 0.14)
Scipio Capicius (1605)	0.69 (0.50 + 0.19)	0.94 (0.75 + 0.19)	1.56 (0.94 + 0.62)

---

[1] Allow, however, for undiscovered homs; see preface.

FAUSTUS ANDROLINUS

Eclogues (1311 lines)

	A		B	
α^1	–		1	0.07
δ^1	7	0.53	4	0.30

α^1 B

4.87	ignea, qui *liquidum pontum* terramque rotundam

δ^1 A

1.60	da, precor, *optata da* tanti numinis aura
4.74	Musae, felices *Musae, quae* pectora semper
6.28	si labor esset iners, *sterili si* litus aratum
10.12	ille furit nec habet *denso quo* possit ab hoste
12.161	*his dictis* mens est iusto suffusa rubore
12.264	tunc ero, *terrestri si* quisquam est orbe, beatus
12.269	iurgia. *di superi,* quanta est dixisse voluptas

δ^1 B

8.19	nec *quisquam tam* laetus erat per compita festa
10.55	et pulsa *optatam iam* paupertate quietem
12.12	frigora *tum primum* gelido nascentia vento
12.30	interea magno *nigrum cum* murmure caelum

ARNOLLETUS

Eclogues (471 lines)

	A		B	
α^1	9	1.89	2	0.42
α^3	1	0.21	–	
β^1	1	0.21	–	
γ^2	1	0.21	–	
δ^1	4	0.85	3	0.64
ε	1	0.21	–	
ζ	3	0.64	3	0.64

α¹ A

1.64	virtutum *magnus cumulus,* celeberrima gesta
1.96	sicine *communes aedes* et iugera linquis
2.31	*tantae maestitiae* questus quis pellere possit
2.44	*internis animis* sane nos ipse colebat
3.28	*fatidicus corvus,* si mens non laeva fuisset
4.28	India *lanigeris lucis* se tollit ad astra
4.68	tunc celeres *diris armis* innexuit alas
4.126	monstra; sed haud *clava praedura* tale subegit
4.145	curabamque *greges pingues,* et sedibus hisce

α¹ B

4.5	*monstrum terrificum* quod demetit omnia falce
4.128	*monstrum terrificum;* Mors ultima linea rerum

α³ A

3.62	atque *dato dono* Persarum more saluta

β¹ A

1.133	ante *tuas aras* pingues mactabimus agnos

γ² A

2.6	heu *crucior vexor laedor* sternorque coquorque

δ¹ A

1.31	interea *quae frondicomae* patet arboris umbra
3.50	sed *qua doctrina* cincti. numeraveris omnes
4.74	*quae durae* valeat praecludere limina Morti
4.90	ventripotens *molli qui* nutrimenta palato

δ¹ B

4.57	haec (hei) commiscet *clarum cum* paupere regem
4.62	o *utinam tam* dira foret demersa sub undis
4.94	o *turpem quem* segnities damnosa moratur

ε A

3.23	at gravior *de te* questus mihi mole malorum

ζ A

1.4	te doceam qua lege *regas teneras* animantes
1.69	cuius *praesidio morbo* plerique levantur
2.2	sicine *pastores gaudes* vexare dolore

ζB

2.11	maerorem *dudum conceptum* fortius auget
3.30	cum *pridem canerem* tenui dictamina culmo
4.146	cottidie *mulctram complebam* lacte recenti

BUCHANAN

	A		B	
De sphaera (c. 2000)				
α^1	16	0.80	2	0.10
α^2	1	0.05	2	0.10
α^3	1	0.05	–	
α^5	1	0.05	–	
α^7	–		1	0.05
β^2	1	0.05	–	
γ^1	2	0.10	–	
γ^3	–		1	0.05
δ^1	3	0.15	4	0.20
ζ	7	0.35	3	0.15
Franciscanus (c. 900)				
α^1	8	0.89	–	
α^3	3	0.33	–	
γ^1	1	0.11	1	0.11
δ^1	3	0.33	3	0.33
ε	3	0.33	–	
ζ	9	1.00	–	

De sphaera

α^1 A

I

sufficiunt *avidis animis:* qua panditur orbis[2]
et trepidum *subitis tenebris* confunderet orbem
Hesperiae metae, longe distaret Eoa
nunc etiam *exiguis spatiis* mensura deesset

---

[2] In the edition available to me the verses are not numbered.

II

donec *equos fessos* Atlantide serus in unda
lumina, non *spatio vasto* semotus Olympus
abdidit, aut *spatiis longis* natura removit
dissita tam *longo spatio* dum sidera spectat
perque *cavas fossas,* declivi tramite, matris
non damna *innumeris saeclis,* non augmina sentit

III

in *partes veteres* ter divisere quaternas
hunc medium *spatiis aequis* utrimque diremptum
non *pluvios Austros,* modo sese inclinat ad Eurum
sed *geminas formas:* nam luces noctibus aequans
vix *tenebras densas* vincat splendore maligno
saevit: et *horriferis flabris* sua regna flagellat

$\alpha^1 B$

I

mane per *adversum clivum* Sol scanderet horis
non satis ad *certam trutinam* coeli exigit orbes

$\alpha^2 A$

I

nam *tenebris primis nobis* cum luna laboret

III

bis *senis faculis* splendescens linea iungit

$\alpha^2 B$

II

victor habet, *quintum gyrum* vix ultimus implet
esse *aliquam primam* vim naturamque necesse est

$\alpha^3 A$

III

dimensis spatiis sic angulus exeat aequus

$$\alpha^5\,\mathrm{A}$$

$$\mathrm{III}$$

alter ab *aequatis tenebris* cum lucibus

$$\alpha^7\,\mathrm{B}$$

$$\mathrm{III}$$

postque *operum socium primum* mactare iuvencum

$$\beta^2\,\mathrm{A}$$

$$\mathrm{II}$$

deseruit *nostris vitiis* offensa nocentes

$$\gamma^1\,\mathrm{A}$$

$$\mathrm{I}$$

tardius *his, illis* citius vaga lumina terris
occidat *his, illis* surgat sol, passibus aequis

$$\gamma^3\,\mathrm{B}$$

$$\mathrm{III}$$

postque *operum socium primum* mactare iuvencum

$$\delta^1\,\mathrm{A}$$

$$\mathrm{II}$$

quod si naturae *vi coeli* caerula templa

$$\mathrm{III}$$

circulus, a *nostri qui* semper tangitur orbis
nam reliquos *coeli, qui* signant aethera, gyros

$$\delta^1\,\mathrm{B}$$

$$\mathrm{II}$$

ad varios *hominum* dum sese accommodat usus

pergit in *occasum dum* navis currit in ortum
crescere: *materiam nam* vis contraria mutans

III

cruraque et *intentum cum* telo in Scorpion arcum

ζ A

I

nam *tenebris primis nobis* cum luna laborat
multiplices decies centena in milia passum

II

longius a *Tauro spatio* breviore recedet
naturae *ingenio nemo* persuaserit auctor
at quibus haec *insunt pereunt* perimuntque vicissim
nec tamen *adversis quamvis* obnoxia causis

III

torrida seu *longas aestas* exporrigat horas

ζ B

I

sylvarum cum strage: omnes sic undique partes
dum nihil *occultum, dum* nil sibi linquit inausum

II

continuo *coelum circum* se volvere flexu

Franciscanus

α¹ A

primum ubi *detonso vesano* vertice caudex
invitis oculis lacrymas simulare, boatu
conceptis verbis de somno surgere, mensam
pectoris *arcanas latebras* cognoveris, omnes
commoda, et *effoetis membris* ignara voluptas
pinguiculas viduas, stupidoque imponere vulgo

et *saturo domino* clementior ulcera lambat
quis *rigidos Scotos* gelidi sub vertice coeli

$$\alpha^3\,A$$

nec simul *exuviis positis* Massyla venenum
iam probe *decurso spatio* spectare propinquam
quam ne *discussis tenebris* resipiscere mundus

$$\gamma^1\,A$$

dissimulet, simulet, praesens ubi postulet usus

$$\gamma^1\,B$$

flectere *cornipedem, volucrem* nec ducere visco

$$\delta^1\,A$$

ergo *cave ne te* falso sub nomine mendax
maiorum *a magna* non degenerare culina
non tamen *his cuivis* fas abgannire, nec arcto

$$\delta^1\,B$$

tum facies *rerum, tum vafrum* protea finge
ecce autem *pronum dum* findit puppe Garumnam
ingratum nunc finge *virum cum* pellice noctes

$$\varepsilon\,A$$

ergo *cave ne te* falso sub nomine mendax
vel proceres merito *de te* bene prodere regi
ut quocumque loco, *de re* quacumque patrata

$$\zeta\,A$$

nam *tenebris primis nobis* cum luna laboret
et quoniam, ut *video, morbo* cruciaris eodem
in venerem *accendas, viduas* quae retia captent
iniice, tunc *humeris quemvis* superingere fascem
sic, *dices, rides,* sic molliter oscula iungis
grande *scelus levibus contentus*[3)] plectere poenis
flectere vix *potui populi* compescere voces
nec *precibus: primus* si non patefecerit arcem

[3)] Counted twice.

NAUGERIUS

Lusus (784 lines)

	A		B	
α^1	5	0.64	–	
β^1	1	0.13	–	
δ^1	5	0.64	2	0.26
ζ	2	0.26	–	

α^1 A

19.4	*desertos scopulos* deviaque antra colis
27.1	pascite, oves, *teneras herbas* per pabula laeta
27.48	cum te *succincta tunica* fusisque capillis
44.25	*antiquae Parcae,* niveo queis corpore amictu
44.36	accipite haec *laetis animis,* neu posse moveri

β^1 A

27.59	pulchraque nulla *meis oculis* te praeter habetur

δ^1 A

6.2	fructum aliquem: *has violas* dat tibi sancta Venus
20.29	pro viridi *cytiso, pro* molli graminis herba
22.13	una *meos, quos* et miserata est, novit amores
25.45	nos hic, *qua nitida* pellucens rivulus unda
26.62	*a domina* lato longius ungue fuit

δ^1 B

27.78	sed nos *dum longum* canimus, iam roscida luna
44.89	*tum demum* placida contentus pace quiesces

ζ A

5.4	*continuo certo* deiicit hanc oculo
44.100	*produces: humiles* sudabunt mella genistae

PETRARCA

Africa (1953 lines)

	A		B	
α^1	12	0.61	4	0.20
α^4	–		1	0.05
α^5	1	0.05	–	
α^7	2	0.10	–	
γ^1	1	0.05	–	
γ^2	1	0.05	–	
γ^3	–		1	0.05
δ^1	13	0.67	4	0.20
δ^2	1	0.05	–	
ε	2	0.10	–	
ζ	3	0.15	4	0.20

α^1 A

1.416	scilicet *immenso studio* dum ledere querit
1.479	hic decet *egregios animos,* hic exitus est quem
1.547	quin tu *animas letas* melioraque regna tenentes
2.53	fata videns, *humiles voces* summissaque verba
2.163	preter *Aquas Sextas* – dicunt sic nomine vallem
2.227	tristia *ceruleis Germanis* bella movebit
2.294	forte sub *extremos annos* mundique ruentis
3.259	lumina *suppliciis variis,* sevoque ministros
3.494	dicitur ad *patrios muros* sparsisse bilustri
3.658	dicere que *campis alienis* castra sequentes
3.670	atque *leves calices* arcere a finibus ause
3.747	persequar *eternis odiis,* nec regna tenere

α^1 B

2.476	neve ibi *tantarum rerum* spem pone tuarum
2.539	*ingratam patriam* – piget heu narrare pudetque
2.543	*Fortunam: patriam servatam* perdere noli
3.711	invenit *attonitum socerum,* pariterque vocati

α^4 B

1.458	invenis atque *animum vacuum.* quin ocius ergo

α^5 A

3.217	ad *nos conversos oculos* vultusque tenebant

α⁷ A

1.61	cesserit *auspiciis: solidis* tunc viribus alta
2.445	iste rudes *Latio duro* modulamine Musas

γ¹ A

1.248	*externo, proprio* non plus in milite fidant

γ² A

2.139	hec michi iam *Scauros Drusos* crebrisque Metellos

γ³ B

3.771	*supplicium scelerum* non una mente tulerunt

δ¹ A

1.10	reddite, *vos animos.* tuque, o certissima mundi
1.17	forte etiam *lacrimas, quas* (sic mens fallitur) olim
1.335	teque putem atque *alios, quos* pridem Roma sepultos
1.425	*his dictis* tulit ante gradum, frontemque modestam
1.429	mortales oculos, *alti si* limina mundi
1.448	*has animas* subterque pedes radiantia solis
1.555	hinc amor. *his ipsis* libertas credita quondam
1.591	*Romanas has* esse animas, quibus una tuende
3.217	ad *nos conversos oculos* vultusque tenebant
3.470	et pelagi *medio, quo* fors contraxerat ambas
3.565	*his dictis* riguere animi, pallorque per omnes
3.598	constitit ingemuitque *ferox, mox* magna precatus
3.749	*his dictis* alios eadem iurare coegit

δ¹ B

1.340	vestra autem mors est, *quam vitam* dicitis. at tu
2.181	tu siquidem letus *mecum, dum* bella gerentur
2.419	anxia *sollicitam quam* non opulentia reddet
3.45	tantorum vindex *scelerum? num* fulmina celo

δ² A

3.122	at gravis *hinc illinc* extantia bracchia Libre

ε A

1.441	eximiumque decus –, quam *de te* concipiam spem
1.451	sunt servata Deo. *qui si* te lumine tanto

ζ A

1.402	obsidet, hinc *vise sese* subducere morti
2.266	at nimium *propero: video* par nobile, natum
2.543	*Fortunam:* patriam servatam perdere noli
3.276	maximus in *magno Scipio* notissimus orbe

ζ B

1.493	conciliumque *hominum sociatum* legibus aequis
2.451	in quod *eum studium* non vis pretiumve movebit
3.620	o genus *eximium, dignum* cui secula cunta

POLITIANUS

Sylvae (2357 lines)

	A		B	
α^1	2	0.08	–	
α^2	–		1	0.04
γ^1	–		1	0.04
γ^3	–		3	0.13
δ^1	13	0.55	2	0.08
δ^2	1	0.04	–	
ε	2	0.08	–	
ζ	5	0.23	2	0.08

α^1 A

4.277	Oceani *refluas undas* molemque natantem
4.593	ipse Lyci *nigros oculos* nigrumque capillum

α^2 B

2.125	ille aliam atque *aliam culturam* dulcis agelli

γ^1 B

3.140	utque Rhodos *Solem, Venerem* Paphos atque Cythera

γ^3 B

3.20	*annorum senium* vitamque in saecla propagent
3.134	ingens *tantorum pretium* mihi crede laborum
4.434	sed Tiberim, *dominum rerum* mundique potentem

δ¹ A

1.29	editus ecce *Maro, quo* non felicior alter
1.42	stat dubius, *vastae quae* primum robora sylvae
1.159	quo segetes veniant *campo, quo* sidere tellus
1.291	at tu *quo, nimio* spoliorum et laudis amore
2.211	*horae, quae* coeli portas atque atria servant
3.96	*has epulas?* quodnam ob meritum, pater optime? certe
3.263	ille tamen quaenam ora *sui qui* vultus Achilli
3.290	Tiresiae *magni, qui* quondam Pallada nudam
3.535	cunctaque complexum, *stabili qui* lege gubernet
4.463	mox comes armorum *Fulvi, qui* sanguine partas
4.671	quippe *alios, quos* nec centum sit dicere linguis
4.725	pingit; et *obscuri qui* semina monstrat amoris
4.779	quemque *operi, ni* me tacita experientia fallit

δ¹ B

1.113	nigraque *dum raucum* tremulis evibrat ab alis
4.164	mens prior it *pessum: tum* clausus inaestuat alto

δ² A

2.30	*huc illuc* vanos ostentans purpura fasces

ε A

2.563	*qui si* certa magis permiserit ocia nobis
3.9	concipit egregium, *si qui* mens ardua conscit

ζ A

1.130	sidera; quique *mihi divini* pectoris haeres
3.139	Solis et *Oceani volventi* progener aevo
3.308	quid doleat *Iuno; coelo* quid portet ab alto
3.557	relligio *numeris; quantis* praesagia signis
4.286	in caput isse *retro liquido* pede fluminis undas

ζ B

3.289	inspuit *augurium; baculum* dat deinde potentem
3.357	elatus *rerum, Balium* Xantumque iugales

PONTANUS

Eclogues 1, 2, 5, 6 (1321 lines)[4]

	A		B	
α^1	9	0.68	–	
γ^1	1	0.08	–	
δ^1	3	0.23	1	0.08
δ^2	4	0.30	–	
ζ	2	0.15	–	

α^1

1. Pompa 2.67	Hercli, *superciliis nigris,* candente papilla
1. Pompa 4.98	praestringit *violas albas* et lilia cana
1. Pompa 5.22	*hircosae setae:* tum guttura collaque circum
1. Pompa 5.257	curvantur *geminae sannae,* quarum altera pontum
1. Pompa 6.7	non rixam *cultus thalamus,* non culcitra litem
1. Pompa 6.8	fert, pacem *thalamus cultus,* fert culcitra somnum
1. Pompa 6.30	et *niveis suris* nigrisque Patulcis ocellis
2.61	*immitis Lachesis,* crinemque e vertice vellit
6.68	cedam ego *cariculis sacris* dulcique placentae

γ^1 A

1 Pompa 2.78	circumstant Aequana *hinc, illinc* innuba Amalphis

δ^1 A

1 Pompa 4.45	est quoque *spes agiles* sciat ut tornare catinos
1 Pompa 5.65	*a leva* coniux felici prole Marana
1 Pompa 7.6	Hesperus adveniet, *socii qui* foedera lecti

δ^1 B

1 Pompa 5.188	*altilium dum* scalpit humi sequiturque parentem

δ^2 A

1.20	spiritus alterno *huc illuc* se miscuit ore
1. Pompa 5.161	*hinc illinc* fluit amnis opacaque ripa virescit
1. Pompa 5.174	effluit *hinc illinc* tepidus liquor, adiuvat uxor
2.139	*huc illuc* agitantur et excitus instrepit aer

---

[4] I take this selection from L. Grilli's *Poeti umanisti maggiori* (Città di Castello, 1914).

ζ A

1. Pompa 2.54   quis superat *nymphas; videas* si forte lavantem
6.44   num det fraga *mihi, cerasi* num molle quasillum

MARTIN RAKOVSKÝ

De magistratu publico (2632 lines)

	A		B	
α^1	12	0.45	1	0.04
α^2	6	0.23	4	0.15
α^3	1	0.04	–	
α^5	1	0.04	–	
α^6	1	0.04	–	
β^1	1	0.04	–	
γ^1	2	0.08	–	
γ^2	1	0.04	–	
γ^3	1	0.04	2	0.08
δ^1	16	0.61	2	0.08
ε	1	0.04	–	
ζ	12	0.45	8	0.30

α^1 A

2.17   heu *quantis furiis* vastos ciet ille tumultus
2.410   cum *geminis alis,* postmodo factus homo
2.426   et *geminis alis* bestia prima, leo
2.428   *imperii Graeci* venter imago fuit
3.43   proterit *hostiles acies* qui Marte feroci
3.186   *divino verbo* redditur ille pius
3.419   hinc *variae technae* rerum et stratagemata in hostem
3.578   *vicini vitii* non adeunda via
3.702   in *mediis lacrymis* sollicitoque metu
3.861   post *morbi varii,* scabies ac lumen ademptum
3.923   dum *lepidis diverticulis* vicibusque stupendis
3.1116   *thesauros magnos* sic satis esse mihi

α^1 B

1.375   namque *bonum tantum* sese communicat ipsum

α^2A

2.601	ordine post *alii multi* variique secuti
2.727	ac *alii multi,* celebrat quos fama per orbem
3.772	et *coeptis cunctis* temporibusque comes
3.1089	sic *oculos multos,* multas auresque manusque
3.1096	*thesauros aliquos* regibus esse decet
3.1184	si *noxas aliquas* dissimulare queat

α^2B

2.81	*omnem splendorem* ponebant atque decorem
2.425	*imperium primum* vertex notat ille gigantis
2.499	*imperium quartum,* contraria bestia votis
2.655	ut genus *alituum multorum* tempore veris

α^3A

| 3.551 | *posthabito Iusto* magnos dant regna tumultus |

α^5A

| 2.705 | ut taceam *proceres tangentes* nomine caelum |

α^6A

| 2.537 | ergo deo *genitus factus* pro sontibus est sons |

β^1A

| 2.574 | cumque *tuis natis* abbreviate Deci |

γ^1A

| 2.76 | *hic, illic* viles ante fuere casae |
| 2.375 | aenea *cesserunt, redierunt* ferrea mundo |

γ^2A

| 3.286 | *res pecudes homines* protinus ure neca |

γ^3A

| 1.264 | *autoris gentis* Romulidumque patris |

γ^3B

| 1.139 | cogat ad *obsequium legum* patris omnipotentis |
| 3.388 | *virtutum reliquum* quae regis una chorum |

δ¹ A

1.18	atque animo *a terra* celsa sub astra vehi
1.35	illas sum procul *a patria* per longa secutus
1.67	sed proceres regesque *tuos, quos* legibus aequis
1.195	nunc ex *qua causa* veniat, quo fonte potestas
3.264	*pro populo* multi tunc gemuere pii
3.337	territus *his dictis,* non culpa, talia supplex
3.408	*pro populo* nec non commoditate loci
3.485	ergo *si iusti* procedere tramite recto
3.517	quis, quid, ubi, quantum, *quo pacto* perspice cuncta
3.518	invenies *animo pro* sapiente scopum
3.567	*pro pulchro* quoties oritur certamen honesto
3.583	haec extrema *cave, ne* vel te audacia perdat
3.1020	rex et in *Aegypto pro* Pharaone fuit
3.1088	quos amat, *hos dominos* optat habere diu

δ¹ B

1.7	*iam linquam* terras et vilem pulvere ludum
1.360	*quam vitam* Princeps monstrat origo, parat

ε A

3.342	ante meos *ne me* dedecorato, precor

ζ A

1.156	Marte *ducis, pacis* tempore regis, obit
2.365	hinc levis *ambitio toto* dedit orbe tumultus
2.587	his subeunt: *nati, Christi* Maxentius hostis
2.615	*sprevisti capiti* promissam intrare coronam
2.692	*dispersos, Henetos* Paphlagonasque viros
3.3	da cytharam, *dixi, plectri* maioris, Apollo
3.68	*suspicimus, cuius* nos simulacra manent
3.339	*peccavi, populi* metuebam namque furorem
3.776	hic *illi caeli* motus ab arce venit
3.875	unde *animae vae vae* subitis horroribus edunt
3.1065	*illi munifici* non sunt largive vocandi
3.1209	rem quamvis *homini multi,* pullumve canemve

ζ B

1.187	esset *dicendum, quantum* discedat ab illa
1.189	*nimirum quantum caelorum*[5] distat ab orbe

---

[5] Counted as two.

1.471	ut sunt *artifices, cives,* famuli atque ministri
2.332	cui *solium fratrum* sanguine sparsit, Ochus
3.155	ultorem *scelerum, castum* verique parentem
3.425	aut qualis *Siculam quandam* perterruit urbem
3.473	*scutigerum quantum* praecellet scriba togatus

SANNAZARIUS

	A		B	
De partu Virginis (1443 lines)				
α^1	2	0.14	–	
α^7	2	0.14	–	
δ^1	14	0.97	1	0.07
ζ	3	0.21	1	0.07
Elegies (1506 lines)				
α^1	3	0.20	–	
γ^1		–	1	0.07
γ^3		–	1	0.07
δ^1	11	0.73	3	0.20
ζ	1	0.07	–	
Others (1388 lines)				
α^1	3	0.22	–	
α^2	1	0.07	–	
α^7	2	0.14	–	
γ^1	3	0.22	–	
δ^1	12	0.89	2	0.14
δ^2	3	0.22	–	

De partu Virginis

α^1 A

2.161	Cappadocum *medios populos* discriminat Iris
3.128	perlustrans *tacitis oculis* loca, concutit alas

α^7 A

1.326	everti *palmas, altas* ad sidera palmas
2.116	*interea terra parta* iam pace marique

δ^1 A

1.43	ut *quos victuros* semper, superisque crearam
1.177	*his dictis,* Regina oculos ad sidera tollens
1.193	dat sese, miscetque *utero, quo* tacta repente
1.357	ast ego *pro nato, pro* te dominoque Deoque
2.162	*praeterea qua* se Thracum Mavortia tellus
2.247	palmiferis Solymas *a laeva* liquerat arces
2.303	expediam: *vos secretos* per devia calleis
2.304	caelicolae, vos *(si merui)* monstrate recessus
3.6	praeterea *quos Eoos* Aurora per ortus
3.90	*quo subito* veteres deponant pectoris iras
3.109	quumque propinquasset *portae quae* maxima caelo
3.364	excutient *oculi, qui* numquam sidera, numquam
3.451	arcana, *hos primos* per signa ostendet honores
3.499	*insolitos: quos* pulchrae udis nevere sub antris

δ^1 B

2.153	quae nauta, *angustum dum* praeterit Hellespontum

ζ A

1.176	usque *adeo magno* nil non superabile caelo est
2.116	*interea terra parta* iam pace marique
3.416	illa autem *humanis quamvis* latura ruinis

ζ B

1.257	Aethiopes, *hominum sanctum* genus, astra secuti

Elegies

α^1 A

1.11.33	quique tot *egregias animas,* tot clara virorum
1.11.36	tamque *pias ferulas* regia sceptra vocet
1.11.55	at vos *incautis sociis* aconita parate

γ^1 B

3.2.13	mille tori *Dryadum, Satyrorum* mille recessus

γ^3 B

3.1.137	praeterea *quantum populorum* Mosa coercet

δ¹A

1.1.6	nunc Liris *gelida qua* fluit amnis aqua
1.7.33	miremur, tibi *si placidi* penetrale Tonantis
1.11.4	desine, *pro populo* stat deus ipse suo
1.11.73	*di patrii,* quorum monitis huc advena classis
1.11.76	moenia *felici si* posuistis ave
2.1.67	gratia *dis Italis;* praerupta Ceraunia, et arces
2.1.115	quare *si nostri* veniet tibi nuntia leti
2.9.26	vanus et *a longa* posteritate ferar
3.2.9	*semiferi, si* vera canunt, domus horrida Fauni
3.3.21	protinus *a dextra* sacrae, mea turba, sorores
3.3.23	*a laeva* nitidis stratum Pythona sagittis

δ¹B

1.3.30	*tam lentam* Lachesis scindat avara colum
1.9.80	Aegidius, *verum dum* canit ore Deum
2.2.41	tu quoque, *quem iuvenem* veneror, dulcissime Garlon

ζA

1.1.3	et *Sinuessanas spectas,* mea gaudia, Nymphas

Other Poems

α¹A

Ecl. 2.68	et videam *nigros populos* Solemque propinquum
Sal. 73	ah *pavidas Nymphas* subitoque horrore rigentes
Epigr. 3.7.9	det fesso *aestivas umbras* sopor: et levis aura

α²B

Fragm. 34	nunc *alios sucos* aliosque Melampodis herbas

α⁷A

Ecl. 3.91	polypus in *scopulis, mediis* melanurus in undis
Mort. Chr. 31	excitasque *umbras medias* ululasse per urbes

γ¹A

Epigr. 1.55.1	omnia *vincebas: sperabas* omnia, Caesar
Fragm. 14	scilicet *Aenaria, Prochyta* venit Euthycus alta
Sal. 50	nullae hic *insidiae, nullae* per aperta latebrae

δ^1 A

Ecl. 1.20	his *oculis, his,* inquam, oculis quae funera vidi
Ecl. 1.57	scilicet *hos thalamos, hos* felices hymenaeos
Ecl. 2.18	at non *Praxinoe me* quondam, non Polybotae
Ecl. 3.2	bis *senos vos,* Mopse, dies tenuere procellae
Ecl. 3.19	nam bene *si memini,* Rhodanum referebat Amilcon
Ecl. 4.9	pro dulci *Latio, pro* nostris detinet arvis
Sal. 57	*his dictis* permulsi animi, securaque tristem
Epigr. 1.4.2	hospes et hunc *acri qui* sedet altus equo
Epigr. 1.13.6	mallet et *hos numeros* quam meminisse suos
Epigr. 2.4.11	si quod *dis genitis* vix tot peperere labores
Epigr. 2.49.4	illa dedit, tristes Actius *hos titulos*[6]
Epigr. 2.54.1	unde mihi *has* violas affers, puer? accipe, mater

δ^1 B

Ecl. 1.76	quo numquam terras *videam iam* iam illa tot annis
Epigr. 1.36.1	Alfonsus *magnum dum* traicit Apenninum

δ^2 A

Ecl. 1.13	*huc illuc,* dum Pausilypi latus omne pererro
Ecl. 5.50	quae ventis agitata *huc illuc* concita fertur
Epigr. 2.52.1	quaeritat *huc illuc* raptum sibi Cypria natum

SCIPIO CAPICIUS

De vate maximo (1605 lines)

	A		B	
α^1	8	0.50	3	0.19
α^3	2	0.12	–	
α^4	1	0.06	–	
α^5	1	0.06	–	
α^7	1	0.06	–	
β^1	1	0.06	–	
γ^3	–		1	0.06
δ^1	11	0.69	7	0.44
ζ	3	0.19	2	0.12

[6] Repeated in the next line.

α^1 A

1.55	impellit *variis studiis* impendere curam
1.98	dispulit *obiectas undas* atque intima vasti
2.156	res nova per *latos populos* urbesque propinquas
2.516	ad *superas auras* revocati et sidera rursus
3.29	namque is *fraternos thalamos* violare, torique
3.70	incensum *infandis odiis* et coniugis ira
3.294	regia *suspensis aulaeis* tota superbis
3.295	splendet et *inductis niveis* mantilibus ingens

α^1 B

1.128	*felicem rorem quem* sudavere tepentes
2.126	*insignem iuvenem* saltus errare per altos
3.442	spectaclo *immanem satiantem* lumina tali

α^3 A

| 2.524 | nec minus et *reliquis defixis* lumina in uno |
| 3.384 | an meritas *sancto fuso* det sanguine poenas |

α^4 A

| 2.411 | spiritus est *alius divinus,* sed tribus una |

α^5 A

| 3.55 | inque ipsa *accensas furias* sedaret amantis |

α^7 A

| 3.24 | divitis *imperii, patrii* quoque nominis heres |

β^1 A

| 2.308 | victa *tuo fluvio* subdunt se marmora ponti |

γ^3 B

| 2.294 | mortalis *rerum dominum* caelique potentem |

δ^1 A

1.170	supplicibus *votis dis vanis* tura dedistis
1.336	hic et supremo *regi qui* sanguine culpam
1.351	at tu magna fide haud *dubia da* dicta probari
1.375	exspectans lucem *quae linguae* frena relaxet
1.493	praevius ostendit *veri qui* lumina solis
2.42	firmabat teneros, *aevi qui* signa futuri

2.178   quae tulit in lucem *nudos, vos* frigida nudos
2.297   elue, nativoque *illic sic* redde nitori
2.507   per me fulgorem *caeli qui* luce carebant
3.104   *his dictis* sese in mollis demisit amantis
3.436   o tantum *his oculis* largus ne tristibus humor

δ¹ B

1.30   *quam patriam* legit, cum caelo missus ab alto
1.136   orbe quaterdeno, *dum natum* perficit aureus
1.149   divinum *imperium tum* flumen sensit et undas
2.342   divinam inspirans *animam, quam* noscere fas est
2.372   nosse, *suum tum* velle oritur, tum intelligit illa
3.255   *verum dum* sanctos monitus metuendaque summi

ε A

3.84   haec *de te* merui? nobis haec praemia tandem
3.88   pollicitis fraudas et *spe me* pascis inani

ζ A

3.319   induit *insertis, vivis* distincta figuris
3.370   ante diem *exstingui crudeli* funere vitam
3.429   felix quum *tali praecingi* tempora fronde

ζ B

1.128   *felicem rorem quem* sudavere tepentes
3.368   aegra *traham dubiam* semper? nostrisque ferocem

BIBLIOTHECA TEUBNERIANA

Anthologia Latina

Rec. D. R. Shackleton Bailey
Pars I. Carmina in codicibus scripta
Fasc. 1. Libri Salmasiani aliorumque carmina
1982. XVI, 382 pages. BT 1030. Cloth DM 125,–

„… das eigentliche Signum der neuen Teubneriana sind SBs eigene Vorschläge zu Wortlaut und Interpunktion der AL. Häufig genug wird an Stellen geändert, an denen bisher keinerlei Anstoß genommen worden war, und erstaunlich oft ist es dabei und in anderen Fällen SB selbst, der eine schlagende Lösung gefunden hat. … In der Geschichte der Beschäftigung mit der AL ist das Buch ein Markstein."

Wolfgang Dieter Lebek, CLASSICAL REVIEW

Cicero
Epistulae ad Atticum

Ed. D. R. Shackleton Bailey
Vol. I: Libri I – VIII
1987. XVIII, 327 pages. BT 1208. Cloth DM 115,–
Vol. II: Libri IX – XVI
1987. IV, 428 pages. BT 1209. Cloth DM 145,–

Cicero
Epistulae ad familiares libri I – XVI

Ed. D. R. Shackleton Bailey
1988. X, 646 pages. BT 1210. Cloth DM 225,–

Cicero
Epistulae ad Quintum fratrem. Epistulae ad M. Brutum.
Commentariolum petitiones. Fragmenta epistularum

Ed. D. R. Shackleton Bailey
1988. VI, 202 pages. BT 1211. Cloth DM 64,–

The revision of the text and apparatus criticus of Shackleton Bailey's Cambridge edition of Cicero's correspondence, in respect of which he was accorded the Charles J. Goodwin award of merit by the American Philological Association and the Kenyon Medal of the British Academy, takes account of subsequently published work, including W. S. Watt's Oxford Texts of *Ad Atticum* I–IV and *Ad familiares* and J. Beaujeu's two volumes in the Budé series. The text has been changed in about 150 places. The last volume of this complete edition also includes the *Commentariolum petitionis* and the fragments, not in the Cambridge edition.

Horatius
Opera

Ed. D. R. Shackleton Bailey
X, 372 Pages
BT 1432. Paperback. Ed. altera 1991. DM 36,–
BT 1437. 1985. Cloth DM 64,–

"This is an important and original work that should interest all Latin specialists and be included in every classical library, ... S. B. makes over 100 proposals of his own, of which he puts over 40 in the text ..."

R. G. M. Nisbet, THE CLASSICAL REVIEW

„... die neue Teubneriana ist eine meisterhafte Leistung, schon weil hier endlich festgefahrene Geleise verlassen werden, ohne daß der Wagen wie bei vergleichbaren früheren Experimenten in den Abgrund stürzt ... Durch seinen immer wachen Spürsinn und seinen Mut, eingewurzelte Ansichten anzugreifen, hat S. B. die wissenschaftliche Arbeit an Horaz auf eine neue Grundlage gestellt. Bentley und Housman würden ihm gratulieren, und ich glaube, auch Horatius Flaccus wäre seinem Herausgeber dankbar, daß er einiges von dem Mißgeschick, das seinen Worten im Laufe der Jahrhunderte zugestoßen ist, wiedergutgemacht hat."

Josef Delz, GNOMON

Lucanus
De bello civili libri X

Ed. D. R. Shackleton Bailey
1988. XII, 321 pages. BT 1502. Cloth DM 116,–

Martialis.
Epigrammata

Ed. D. R. Shackleton Bailey
1990. XX, 542 pages. BT 1531. Cloth DM 168,–

"... this is a fresh edition, greatly superior to its predecessors, and it should find a place in every classical library. ... The new edition, as was to be expected, makes full use of Housman, whose clear-headedness is shown here at its most striking ... S. B. has also incorporated his own notable contributions to the text and interpretation of Martial, and he has added a fair number of new ideas. Above all he shows exemplary judgement in his choice of manuscript variants: one feels for the first time that one is reading a text of Martial that can regularly be expected to make sense. ... It remains to congratulate Professor Shackleton Bailey once more on another outstanding edition."

R. G. M. Nisbet, THE CLASSICAL REVIEW

Quintilianus
Declamationes minores

Ed. D. R. Shackleton Bailey
1989. VI, 425 pages. BT 1753. Cloth DM 148,–

Onomasticon to Cicero's Speeches

By D. R. Shackleton Bailey
Second revised edition. 1991. 152 pages. Cloth DM 58,–

"In the century and a half since the only attempt at a complete onomasticon to Cicero, our knowledge of the people and of the period has grown tremendously, and the texts themselves have undergone significant recension. This *Onomasticon to Cicero's Speeches*, keyed to the standard reference works, thus fills a long-standing need. Furthermore, in the compilation of citations for individuals, the author's inclusion of references in which the individuals are not actually mentioned by name makes this an even more useful tool than one could hope. Finally, the author has done a great deal more than render this material more accessible than it is at present. His many comments upon individual items and his note on *tria nomina* are based on a remarkable knowledge of both the text and the history of the period, and are worthy of publication in themselves. In shore, this book ist both a useful tool and a significant contribution; it will become a standard reference work in the field."

Christopher Craig, University of Tennessee

In preparation:

Onomasticon to Cicero's letters
1995

Abbreviations – Tria Nomina – Persons and Deities – Cognomina – Places – Laws – The thirty-five tribes – Miscellaneous – Headings – Quotations – Appendix: H. Solin's review

B. G. TEUBNER STUTTGART UND LEIPZIG

B. G. TEUBNER STUTTGART UND LEIPZIG